SCIENTISTS
IN EVERY
BOARDROOM

Harnessing the Power of
STEMM Leaders in an Irrational World

RUBY CAMPBELL, PHD

Praise for Scientists in Every Boardroom

"Now more than ever does business need rigorous, evidence-based thinking to guide them and society through the crisis of ideologically driven drivel that passes for much political and economic thinking from our supposed leaders. In an accessible, informed fashion, this book with help business do just that. Highly recommended."

Professor Anthony M Grant, Director, Coaching Psychology Unit, University of Sydney

"This is an important, practical book and one that I would recommend every board puts on its reading list. Fake news is not just confined to the political daily news. Counter-factual information (counter-knowledge) in organisations causes immense damage, results in bad decisions, is costly, and is more common than many organisations would wish to admit. Ruby Campbell's call for greater evidence-based practice is not only timely but one that needs to be taken seriously in all organisations. The lessons and techniques contained in this book should be written on every organisation's walls."

David Wilkinson, Chief Editor, *The Oxford Review*

"To secure Australia's health and economy into the future, the leadership of women and men in STEMM is vital. *Scientists in Every Boardroom* provides much-needed inspiration for all professionals working in science, technology, engineering, mathematics and medicine."

Rozalina Schulze, Founder and CEO, Health Industry Hub

"The need for systemic thinking and leadership to address the world's 'wicked' problems has never been greater. Ruby Campbell has hit the proverbial 'nail on the head': declining trust in traditional institutions, a world crying out for rational and ethical leadership and an impending tidal wave ready to sweep us away if we don't change. More than that, however, Ruby has set out a cohesive and readily understood way to *address* the issues. This book is an investment well worth making!"

Chris Whelan, Director, The Transformation Initiative

"Such an insightful read into what leadership needs to evolve into. Covering multiple and often forgotten elements of what it will take to be a successful leader for the future, this book clearly details the logical steps to transition into one while providing scientific evidence and data in an easy-to-read format to support the concepts. This book is a must-read for anyone looking to make a genuine difference moving forwards."

John Poulter, Director, Comstor Australia

"As Dr Ruby Campbell says, 'the world urgently needs more critical thinking'. This book goes beyond talking about what will drive a profit for business but considers the type of leaders needed to deliver better outcomes for humans on this Earth. Both inspirational and useful, Ruby's thinking will help individuals, organisations and nations achieve better results, for this and future generations, through a new wave of leadership."

Jen Marshall, Chief Product Officer, Isentia Australia

"Tired of the litany of poor, shortsighted decisions that pass for leadership and exacerbate navigating our increasingly complex world? Dr Ruby Campbell offers accessible evidence and anecdotes for the power and possibility that scientific thinking can provide, especially when applied to society's wicked problems. More than ever, we need the elevated, creative head (and heart and guts) to make our world workable. Though not *the* answer, it is *an* answer to what can enable leaders to lead. A worthy read."

John B. Lazar, Leadership/Executive Coach, CEO/Founder,
John B. Lazar & Associates, Inc.

"*Scientists in Every Boardroom* is insightful in understanding our current leadership environment, and how, with the right coaching, people who have traditionally taken a backseat can reach their leadership potential, and impact your business positively."

Clive Lam, Product Director, Protiviti, Hong Kong

"A much-needed and timely reminder that quality and well-credentialled scientists are required for business and community leadership in making informed decisions and implementing them, based upon scientific principles, education and disciplines, in both meanings."

David Edmonds, Royal Australian Chemical Institute

To my son Isaak,

My love for you, like your courage, is boundless.

First published in 2019 by Ruby Campbell

© Ruby Campbell 2019
The moral rights of the author have been asserted

All inquiries should be made to the author.

A catalogue entry for this book is available from the National Library of Australia.

ISBN: 978-1-925921-37-3

Project management and text design by Michael Hanrahan Publishing
Cover design by Peter Reardon

Disclaimer
The material in this publication is of the nature of general comment only and does
not represent professional advice. It is not intended to provide specific guidance for
particular circumstances and it should not be relied on as the basis for any decision
to take action or not take action on any matter which it covers. Readers should
obtain professional advice where appropriate, before making any such decision.
To the maximum extent permitted by law, the author and publisher disclaim all
responsibility and liability to any person, arising directly or indirectly from any
person taking or not taking action based on the information in this publication.

CONTENTS

AUTHOR'S NOTE

I started writing this book in 2016, however at that time it was going to be about *coaching for ethical leadership*. Quite unexpectedly, the research process, coupled with world events, drew my attention to what has become a life mission: to help STEMM (science, technology, engineering, maths and medicine) professionals reach the top. My passion for coaching STEMM leaders to transition to the boardroom and C-suite comes from both success and failure. In a way, I've been preparing for this book my whole life.

I'm profoundly grateful for my life experiences: a mix of exciting and often challenging events, having relocated to Australia from Puerto Rico with my family in 1979, when I was a young girl. My initial passion, what I loved doing throughout my childhood, was writing and exercising the power of communication. However, that was in Spanish, which is the official language in Puerto Rico. Here I was at age 15, in a new country, unable to speak or write English fluently. For a while I felt like a mute. However, I didn't have the luxury of time to become proficient enough in English to pursue a career in writing. I therefore looked for another area that brought me great joy.

My dad was a voracious reader about all kinds of topics: politics, geography, cinematography, literature … science. He was a walking encyclopedia and I wanted to be just like him. We used to spend hours in the evenings talking about what soon became our common ground: science. He was adamant that I was best suited to a career in science, engineering or medicine. I faint at the sight of blood and my spatial skills aren't that great. I chose science; chemistry, to be specific. To me, chemistry seemed to explain pretty much how everything worked.

I was fascinated by the lives and achievements of brilliant scientists, engineers and physicians such as Albert Einstein, Nikola Tesla, Dmitri Mendeleev, Rosalind Franklin, Marie Curie, Louis Pasteur, Michael Faraday, Thomas Edison and Henry Ford, among many others. Reading about these giants of knowledge at my favourite spot in the library, I'd be so engrossed in their stories of relentless enquiry of the unknown, challenging the "known", their commitment to discovery and innovation, and other-worldly persistence and tenacity that I'd often forget to eat my lunch. Their stories spoke directly to my soul – or the mind and consciousness, according to the neuroscientific perspective. Their message was that scientists transform the world with their discoveries, and that they're innately driven by curiosity and, in most cases, by a desire to do good. It was in that library, two years prior to finishing high school, that I decided I wanted to be a chemist and somehow contribute to society.

Over time, I decided on a career in the pharmaceutical industry; developing new medicines to improve public health became my higher purpose. Of course, back in the 1980s, choosing a career in science was not the norm for women in Australia. That's not to say that pursuing a career in science, engineering or medical sciences was popular, much less considered cool, for boys in those days either.

With great enthusiasm and dedication, I channelled my innate curiosity into a scientific career in chemistry, becoming one of the first female R&D managers in the Australian pharmaceutical industry in the 1990s, followed by a rapid rise during my late 20s to mid-30s to leadership roles both nationally and internationally. On the surface, I appeared to have it all, to be a trailblazer, on my way to shattering the glass ceiling in the male-dominated STEMM world. However, signs of my ineffective coping mechanisms for dealing with stress (at levels that nowadays are considered untenable) started to show, both physically and emotionally. This was the 1990s and early 2000s, when "working hard and playing hard" were the order of the day, and when women working in STEMM were not the norm, much less in corporate leadership roles and with a PhD under their belt.

It wasn't until I hired an executive coach that I began to "hold up a mirror". The emotional and cognitive scaffolding generously afforded by my coach helped me come to a pivotal realisation: having a PhD in Science and doing conventional skills training courses (both in-house and external) are, in most cases, significantly insufficient preparation for a senior leadership role in the 21st century. Thirsty for deeper knowledge and insight into what constituted a great leader, I undertook an Executive MBA at one of Australia's most respected business schools. The experience was transformative, to say the least. It exposed me to fellow directors and general managers from other fields such as financial services and management consulting. It was those interactions that made me question whether I had truly cracked "the leadership code". I kept wondering if we were all missing something, especially since we seemed to gloss over the real reasons behind the leadership crisis leading to the Global Financial Crisis (which had unfolded before our eyes as business students). After a heartfelt discussion with the Executive Director of Executive Education, I enrolled in the world's preeminent Masters of Coaching Psychology at The University of Sydney, straight after completing my Executive MBA. (I may have a masochistic penchant for tertiary education.)

Equipped with a new lexicon and a scientific and practical understanding of human and organisational behaviour, I was able to look back and make sense of my career trajectory so far. For the first time in decades, I was filled with hope and excitement about the future. Soon I became aware that the world was ready for a different approach to leadership development. It's been almost 10 years since I started coaching leaders, as the Founder and Managing Director of a learning and development (L&D) consulting firm dedicated to growing the right leaders and organisations in the 21st century. It is with a great sense of joy, humility and privilege that I turn my attention to researching and coaching STEMM leaders, for reasons that will hopefully become as compelling to the reader as they are to me.

This book is the culmination of three decades of research and experience in leadership, business, science and psychology, as I gratefully stand on the shoulders of other executive coaches, psychologists, leadership experts and journalists. It is my hope that the reader comes away with a greater understanding of why society needs more leaders from STEMM backgrounds, the many challenges (often in areas typically associated with minority groups) faced by STEMM professionals, and finally, with a roadmap for STEMM leaders to transition to the C-suite and other influential positions.

INTRODUCTION

A WORLDWIDE CRISIS IN RATIONAL LEADERSHIP

NOW, MORE THAN EVER ...

Now, more than ever, the wellbeing (some might say the survival) of humanity is in the hands of government and business leaders who are making decisions that have become more complex than ever before in human history. We face many global challenges, including climate change, international tensions, technological instabilities, trade wars, increasing mental health problems, failing leadership, drug problems and national political upheaval. The question is: are our current leaders equipped to handle such situations and make the complex decisions needed to address them?

Despite remarkable human progress, global risks seem to be rapidly intensifying. At the time of writing of this book, most Western governments and business leaders seem to be floundering when it comes to rational and responsible decision-making, ignoring scientific evidence and thereby threatening the wellbeing of our societies and the planet. Granted, there are many calls to action in global forums organised by the United Nations and the World Economic Forum, however they often seem to be more talk fests than anything else.

The past 35 years have seen most business and government leaders focusing almost exclusively on financial and economic growth, often at the expense of other measures such as the environment and social issues. Perhaps this is because, until recently, our only measure of a country's prosperity was gross domestic product (GDP). Likewise, for businesses most people have only been interested in ongoing profitability and growth. But we now know that the wellbeing of an individual, an organisation and a country consists of much more than economic wellbeing (although this remains crucially important).

Promoting STEMM professionals to the C-suite or board positions in the highly influential corporate sector will undoubtedly improve the decision-making quality of large companies. But regrettably, the data shows that STEMM leaders aren't often promoted to business leadership positions, unless the company is founded by a STEMM professional. This lack of promotion is largely due to an interplay of individual and organisational cognitive biases, hindering interpersonal and other skills development.

This book seeks to demonstrate how to apply the SCIENCE of Leadership coaching model in helping STEMM leaders develop the necessary leadership capabilities to transition to the boardroom, C-level roles, and beyond. This is done by discussing several key models from coaching science, and by following the stories of two leaders (coaching clients) working in STEMM industries: Gina and Rob. These coaching clients are composites: in the same way a movie is made of thousands of still images/frames edited together to tell a story, these case studies are composed of fragments of dozens of real-life corporate leaders. This technique was used because it protects the identities of our real clients, and partly because this larger narrative is more effective at explaining the science of the coaching methodology than standalone vignettes. The research doesn't come close to addressing *every* leader's experience, however I hope that these stories (case studies) will give you that sense of how each individual's experience is unique and, at the same time, universal.

I hope that business and Human Resources (HR) leaders will be delighted to learn that coaching STEMM leaders with the SCIENCE of Leadership model can lead to tangible and intangible benefits such as greater innovation, employee engagement, profitability and growth.[1] Moreover, in today's world where businesses are committing to the UN's Sustainable Development Goals, which require STEMM leadership front and centre, this represents a win–win–win scenario (that is, it benefits the coaching client, the organisation, and the world).

In this book I will walk you through the unique challenges faced by anyone leading STEMM professionals and how to manage those challenges effectively. If you are a STEMM individual yourself, this book will help you understand why you may have found it difficult to transition to senior leadership roles, or get to the C-suite or the board level – and it will show you how you can change that. If you are a business leader, board member, or senior manager/leader and you are looking to either add STEMM individuals to your team or get the most out of those who are already in your organisation, this book will:

- Help you to better understand the unique qualities of STEMM professionals
- Identify and debunk the myths and biases preventing teamwork
- Demonstrate why it's crucial to have STEMM professionals on your leadership team
- Show you how to develop their leadership capabilities systematically
- Identify the links between the performance and wellbeing of STEMM leaders, with the performance of the organisation, business and the wider society.

The following issues will be examined throughout the book.

---

1   We'll use the term HR throughout the book to denote the evolving function responsible for people and culture within organisations, including but not limited to talent selection and onboarding, employee engagement and performance systems, and learning and development (L&D). HR is also known as People & Culture, and Human Capital.

WHAT IS STEM, AND STEMM?

While the definitions vary, for the purposes of this book, STEM and STEMM may be used interchangeably, as follows:

- Science encompasses disciplines within the natural and physical sciences, and selected disciplines from agriculture and environmental studies: astronomy and the earth sciences, physics, chemistry, materials science, biology and environmental science. These sciences are characterised by systematic observation, critical experimentation, and rigorous testing of hypotheses.

- Technology provides goods and services to satisfy real-world needs, operating at the cross-section of science and society. Information and communications technology (ICT) play an ever-increasing role in society and provides enabling capacity to other STEMM disciplines.

- Engineering draws on scientific, mathematical and technological knowledge and methods to design and implement physical and information-based products, systems and services that address human needs, safely and reliably. Engineering considers economic, environmental and aesthetic factors.

- Mathematics seeks to understand the world by performing symbolic reasoning and computation on abstract structures and patterns in nature. It unearths relationships among these structures and captures certain features of the world through the processes of modelling, formal reasoning and computation.

- Medicine (or medical sciences) is the science or practice of the diagnosis, treatment, and prevention of disease (in technical use often taken to exclude surgery). Contemporary medicine applies biomedical sciences, biomedical research, genetics, and medical technology to diagnose, treat, and prevent injury and disease, typically through pharmaceuticals or surgery, but also through therapies as diverse as psychotherapy, external splints and traction, medical devices, biologics, and ionising radiation, among others.

WHY DO WE NEED STEMM LEADERSHIP?

Science, research and innovation are widely recognised as key to boosting productivity, creating more and better jobs, enhancing competitiveness and growing an economy, as highlighted in the 2016 Government report "Australia's STEM Workforce". Their importance has been accepted in mainstream economic theory for some time.

It's estimated that the advanced sciences (biological, physical and mathematical sciences) directly underpinned around 14% of Australian economic activity in 2012–13. When flow-on effects are considered, the impact of these STEM fields amounted to over 26% of Australian economic activity, or about $330 billion per year.[2]

In the US, scientific and technological advances were estimated to account for approximately 50% of all national growth in the 50 years to 2004.

More importantly, the world has seen overall improvements in quality of life and wellbeing because of STEMM advances. For example, in most parts of the world, humans no longer fight for their lives as a result of infection, thanks to inventions such as penicillin and other antibiotics.

Thinking back on my first role as a scientific senior executive in the corporate world and myriad conversations with STEMM professionals and leaders during my 35-year career (25 years as a senior leader, and 10 years as an executive coach), a few common themes surface. One of them is the frustration of reporting to senior executives who cannot understand a scientific argument to make a balanced and rational decision, and their preferences for quick and often ill-conceived decisions and focus on short-term results. Many corporate C-level leaders (this includes governments and other institutions) are driven primarily by profitability and rapid growth. Decisions are often made without consideration of all stakeholders (not just the shareholders) and long-term implications.

---

2   Australian Government Office of the Chief Scientist (2016). *Australia's STEM Workforce: Science, Technology, Engineering and Mathematics* (pp. 2-158). Canberra, Australia: Commonwealth of Australia.

On the other hand, the decisions that leaders must navigate today are increasingly complex, not to mention having to manage constant pressures from multiple and conflicting stakeholders. The higher up a leader is in the organisation, the harder it is to make balanced decisions. However, if we consider the reasons why somebody wishes to lead, they would ideally need to be linked to the leader's values, and the values of the followers, as suggested by the latest research in motivation. This is especially pertinent when we view leadership as *a process whereby an individual influences a group of individuals to achieve a common goal*. The goals need to be meaningful to everyone; that is, they need to be aligned with our values.

We now understand that to truly gauge how well an individual, an organisation or a country is performing, measures other than simply finances need to be considered – for instance:

- Physical and mental health
- Gainful employment
- Housing access
- Social justice
- Education
- Engagement and governance
- A clean and safe environment.

This leads to the question: shouldn't organisations and governments around the world set goals with the view to achieving wellbeing for their employees and its citizens, respectively and collectively?

SMALL SIGNS OF PROGRESS

It will be crucial for governments, businesses and other international institutions to collaborate in the development of the necessary STEM ecosystem to drive innovation, economic growth and the overall wellbeing of the population in Australia. In varying degrees, the same applies to all other countries. There have recently been some small signs of progress.

The wellbeing budget

At the time of writing, New Zealand's Prime Minister Jacinda Ardern had just unveiled the country's first wellbeing budget, following in the footsteps of Bhutan. Bhutan introduced the concept of Gross National Happiness to the UN in 2011, and since then, the UN has been publishing the "World Happiness Report", applying emerging measures from researchers in the field. These efforts have resulted in changes to the narrative about the relevance of conventional economic growth indicators.

The UN Sustainable Development Goals

Until recently, it was difficult to find a secular, apolitical, rational and responsible approach to achieving goals that would holistically consider the economy, society and the environment. In September 2015, the United Nations convened its member nations and agreed on 17 Sustainable Development Goals (shown in figure 1), also known as the Global Goals, to address the many problems facing humanity and transform the world by 2030. Never have we seen such a coordinated, interconnected and comprehensive approach to tackling the breadth and depth of humanity's problems.

Achieving such a mammoth task requires the best minds in the world, especially those creating, inventing and innovating in science, technology, engineering, mathematics and medicine. However, sometimes it seems like those discoveries that have the potential to truly help humanity do not come to the surface quickly enough, if at all. They seem to get lost in the sea of politics, lobby groups and the media, driven by unmitigated capitalism. It often seems as though STEMM minds are not always empowered to make decisions regarding what's best for humanity. That power has been given to, or taken by, leaders with limited or no STEMM background. And as we'll see throughout the book, quite often it seems like we're being led by irrationality.

Figure 1: The 17 UN Sustainable Development Goals or Global Goals

1	2	3	4	5	6
No poverty	Zero hunger	Good health and well-being	Quality education	Gender equality	Clean water and sanitation

7	8	9	10	11	12
Affordable and clean energy	Decent work and economic growth	Industry innovation and infrastructure	Reduced inequalities	Sustainable cities and communities	Responsible consumption and production

13	14	15	16	17	
Climate action	Life below water	Life on land	Peace, justice and strong institutions	Partnerships for the goals	Sustainable development goals

The National Innovation and Science Agenda

Driven by global attention to the socioeconomic impact of digital technology and AI, the understanding that innovation is the main driver of economic growth, and international education reforms to focus on STEM education, the Australian Government launched the $1.1 billion National Innovation and Science Agenda in 2015. Over $64 million has been allocated to fund early learning and school STEM initiatives under the Inspiring All Australians in Digital Literacy and STEM measure. Schools and universities around the country are promoting and investing in STEM education significantly more than before, and that's something to celebrate. The acronym STEM is used by Australia's Chief Scientist Alan Finkel AO in the 2016 government report "Australia's STEM Workforce". Dr Finkel states that Australia's future will rely on STEM disciplines and, as such, his office embarked on identifying who possesses these skills in our nation and what happens to those individuals once they graduate.

MASSIVE UNTAPPED POTENTIAL

There is tremendous untapped power at the intersection of influential STEMM leadership and the transformation of the world. The 17 Sustainable Development Goals launched by the UN to transform the world by 2030 require scientific and technological innovation and scientific decision-making processes. With more cognitively diverse leadership that includes STEMM leaders, the achievement of the 17 UN Global Goals may become a reality by 2030. Unfortunately, the world has lost trust in institutional leadership such as governments, especially in the UK, the US and Australia. People are now looking to their employers to drive the necessary changes, however this brings us back to the issue of having insufficient board and C-level leaders with a STEMM background.

Neglecting the power of STEMM leaders creates conflict within organisations as STEMM staff feel misunderstood, unfulfilled and disengaged, ultimately leaving the organisation. This is tremendously costly to the organisation, not to mention the unrealised innovation opportunities. Generic learning and development solutions are not delivering the return on investment expected by business leaders. Adopting leadership development coaching programs specifically tailored to STEMM leaders, and providing the right organisational environment, will enable their progression to influential and transformational leadership roles. We must also look beyond the corporate sector and consider having more STEMM leaders in government and other institutions.

There were, and still are, certain stigmas associated with being a STEMM leader, as shown by derogatory terms like "nerd", "geek", "bookworm" and "mad scientist" in Western countries. Such negative words imply personality flaws such as social awkwardness and extreme shyness in people pursuing a career in science, technology, engineering, mathematics and medicine. This is certainly what I experienced studying and working in science in Australia in the 1980s and '90s, and to a lesser degree as I successfully moved up the corporate ladder in the 2000s and beyond. It was far from easy, as countless STEMM professionals will attest. Indeed, in this book we'll see how it can be unnecessarily treacherous for technical experts who are promoted to senior leadership roles without the right development and support.

THE WORLD URGENTLY NEEDS MORE CRITICAL THINKING

This book is not about politics, nor is it about criticising past and current leaders (although at times, it may read as such). Instead, this book is about presenting the results of extensive experience, research, reviews and observations on leadership, in the hope that we'll find some potential solutions to what seems to be a crisis in rational leadership.

More importantly, this book is about empowering STEMM professionals everywhere, as well as their managers and their HR business partners, to pursue influential leadership roles in all areas and make the necessary systemic changes to achieve it. This book is also for leaders in the corporate world, governments and non-profits, to help them understand the world of STEMM and how STEMM leaders can be a powerful force for change and for good. The world urgently needs more critical thinking. We need more rational decisions, not emotion, manipulation, demagoguery and gut feeling. Our brain's autopilot (that is, intuition) can serve us well, but poor thinking and corrupted motivations are behind many of the world's biggest problems.

Scientific minds can help turn the tide. This is a matter of preserving our planet for our children and for future generations.

PART I

WHY WE NEED CHANGE

CHAPTER 1

THE CHANGING LEADERSHIP
NEEDS OF THE 21st CENTURY

*"Trust is the glue of life. It's the most essential ingredient
in effective communication. It's the foundational principle
that holds all relationships."*

Stephen Covey

Despite remarkable human progress, global risks seem to be intensifying and becoming more complex every day, such as climate catastrophes, political instability, technological instabilities, mental health problems and biological threats. The crisis of leadership is extending to almost every aspect of society. Religions, not-for-profits, national governments, local governments, sporting bodies, big business, media outlets – all have been hit with scandals and failures in recent years that have left the public bemused, disillusioned and often angry.

The numbers support this sense of frustration and lack of faith. The World Economic Forum (WEF) released the results of a global survey in 2014 showing that 86% of respondents agreed that the world was in a leadership crisis. And things only seem to have become worse since then. The reasons are as complex as they are varied. The authors speculated it was because the international community had repeatedly failed to address major global issues in recent years. It had failed to deal with global warming, and it barely dealt with the failure of the global economy, causing severe and widespread problems in Europe and North America. Meanwhile, the region often most affected by violence, the Middle East, has been left to worsen in bloodshed.

However, it's not just government leaders who seem to have failed us. Figure 2 shows that, out of a confidence rating of 10, all sectors in the WEF survey rated below 5 except for non-profit organisations. Unsurprisingly, political leadership took the most heat in nearly every country surveyed by the WEF, which led to questioning the effectiveness of current global governance structures such as the UN, WTO, IMF and G20.

Figure 2: Trust in leadership
(rating out of 10, where 10 is the highest level of confidence)

Sector	Trust in leadership
Non-profit and charitable organisations	5.53
Business	4.72
Education	4.70
International organisations	4.62
Healthcare	4.53
News media	3.94
Government	3.83
Religious organisations	3.57

Adapted from: World Economic Forum, Outlook 2015 Report

In 2018, Forbes.com published the article "5 Reasons Why Leadership is in Crisis", listing them to be:

1   leaders focus on outcomes instead of causes

2   leaders believe organisations to be machines; that is, they lack systems thinking

3   leaders fail to see beyond their egos

4   leaders lack self-awareness

5   leaders venerate activity and meaningless achievement.

Leadership experts continue to write about our current leadership crisis, citing lack of trust from the public as the underlying challenge. This is consistent with the WEF 2015 report, which suggests declining trust in our leaders. However, does that imply that the world is getting worse? And why would that decline in trust matter? Let's consider the facts.

IS THE WORLD GETTING BETTER OR WORSE?

The notion that we're experiencing a leadership crisis seems to be in stark contrast with the good news presented by professor of international health Dr Hans Rosling, and co-authors Ola Rosling and Anna Rosling Rönnlund, in their 2018 bestseller *Factfulness: Ten reasons we're wrong about the world – and why things are better than you think*. As the subtitle suggests, the authors believe we are wrong in thinking that things are getting worse, and instead things are actually better than we think. The hard facts show there's less poverty than ever before, people everywhere are living longer, and less of the world is being run by sexist and oppressive patriarchs.

According to the authors, to think that the world is getting worse is "a mega misconception" created by 10 human instincts that trick our brains into perceiving the world inaccurately. To see the world accurately, the authors suggest, people need to take in multiple perspectives and avoid casting blame on individuals or groups. Furthermore, we are cautioned against rash decisions, exaggerations, and "gut feelings", and are encouraged to instead look at the facts, the evidence. This sounds very much like the scientific method and the case for evidence-based decision-making, which is central to the case presented in this book regarding the advancement of more STEMM leaders to influential positions. More on that further on.

What the numbers say

The argument that the world is better now is also made by Harvard Professor Steven Pinker, in his 2019 book *Enlightenment Now: The case for reason, science, humanism, and progress.* In this book (described by Bill Gates as being his "favourite book of all time"), Pinker eloquently explores each dimension of enlightenment as he seeks to demonstrate that, if we dare to understand, progress is possible in all fields: scientific, political and moral. Quoting the 21st century physicist David Deutsch, Pinker opens chapter 1 with the following definition for optimism: "optimism is the theory that all failures, all evils, are due to insufficient knowledge ... Problems are inevitable, because our knowledge will always be infinitely far from complete ... ". In other words, problems can be eventually solved – it's just a matter of acquiring the knowledge needed to formulate a solution.

To access reliable sources of research about the state of the world, you may wish to visit the Our World in Data website. You will discover qualitative and quantitative information on world conditions from 1800 to 2015, looking at specific indicators such as poverty, income inequality, health, child mortality, literacy and hunger. It is closely integrated with the website SDG Tracker, where data and research on the UN's 17 Sustainable Development Goals (SDGs) is presented.

In 2015, all countries in the world agreed to work towards achieving the SDGs by 2030. The SDG Tracker is the only resource that presents all the latest available data on the 232 SDG Indicators by which the 17 goals are assessed. The World in Data and the SDG Tracker are collaborative efforts between researchers at the University of Oxford – who are the scientific editors of the website content – and the non-profit organisation Global Change Data Lab, which publishes and maintains the website and the data tools.

Is the world much better? Is it awful? We must study the data to know all perspectives on global living conditions. By studying the world through data, these facts are impossible to miss. However, the facts on how the world is changing are not known to most people because many of those who report on how the world is changing do not take the data seriously. This needs to change. What we must achieve as leaders is to first understand both perspectives at the same time: we need to know how troubled the world is, and also that a better world is possible. It's ironic that in a world where knowledge and education are improving dramatically, there is abysmal, widespread ignorance about the improving state of the world.

Figure 3: Global child mortality rates

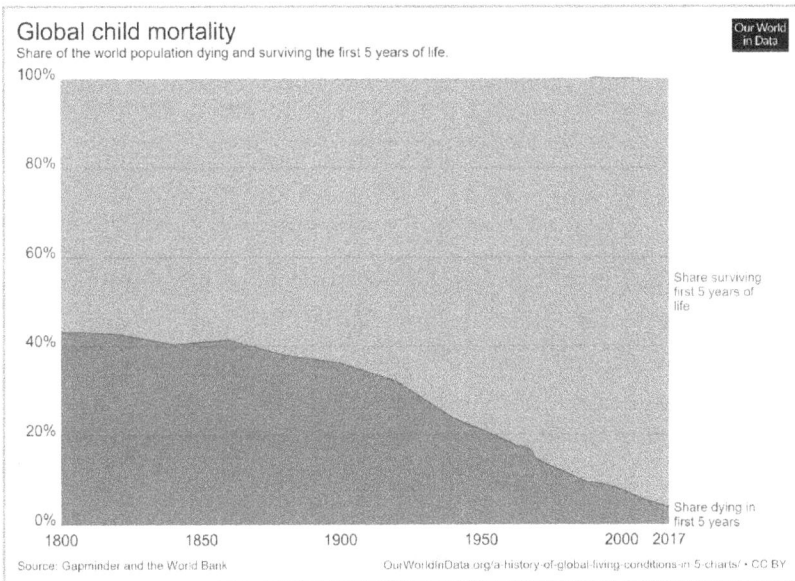

Global child mortality
Share of the world population dying and surviving the first 5 years of life.

Share surviving first 5 years of life

Share dying in first 5 years

Source: Gapminder and the World Bank

OurWorldInData.org/a-history-of-global-living-conditions-in-5-charts/ • CC BY

Researchers at Our World In Data suggest that the media is partly to blame. The media does not tell us how the world is changing; it tells us where the world is going wrong. It tends to focus particularly on single events that have gone bad. By contrast, positive developments happen slowly, which leaves journalists with no event to promote in a headline. "More People Are Healthy Today Than Yesterday" won't sell many papers or get many clicks. The result is that most people are ignorant about how the state of the world has changed. It's easier – and usually more profitable – to scare people than to instil them with confidence.

If we are going to address today's challenges we must know what is wrong with the world, but given the scale of the positive changes we have achieved already and what is possible for the future, it seems irresponsible to only report on how poor our situation is. To make matters worse (for any positive news), the human brain is designed to detect and amplify negative messages relatively easily. What the data tells us is that we know it's possible to make the world a better place because we have already made great progress. There are hundreds of thousands of people currently working on solutions, and there are an infinite number of other ideas yet to be conceived. This is where STEMM leaders come in, with their unshakable determination that science will continue to transform the world.

There is no black and white solution, nor is there a binary way of looking at the world. The fact is, there is still great suffering and peril, and we mustn't stop improving the lives of every human on the planet ("leave no one behind", as the UN exhorts). Millions of children and adults are still going hungry around the world, not to mention the increasing problems linked to human displacement and refugees. Furthermore, there are new and perhaps greater risks in 2019, with a greater impact on humanity than we've ever seen before, as explained by the WEF in their "Global Risks Report 2019" (see figure 4).

Although the lists are distinct, there are common themes that connect the two, with environmental concerns accounting for three of the top five risks in likelihood and four in impact. Our inadequate response to the threats posed by climate change and to reducing greenhouse emissions is in second place on both lists, reflecting concerns about environmental policy failure.

Figure 4: Global risks (WEF 2019)

Top five risks by likelihood	Top five risks by impact
Extreme weather events (floods, storms, etc.)	Weapons of mass destruction
Failure of climate-change mitigation and adaptation	Failure of climate-change mitigation and adaptation
Major natural disasters (earthquake, tsunami, volcanic eruptions, geomagnetic storms)	Extreme weather events (floods, storms, etc.)
Massive incident of data fraud/ theft	Water crisis
Large-scale cyberattacks	Major natural disasters (earthquake, tsunami, volcanic eruptions, geomagnetic storms)

The most dramatic decline in trust in institutions – ever

It's therefore not surprising to learn that 2018 saw the most dramatic decline in trust in institutions ever measured by the Edelman Trust Barometer, both in the US and in Australia.

Lack of trust in governments, businesses, non-governmental organisations and the media is indeed a significant problem here in Australia, as discussed in a recent report by the Commonwealth Scientific and Industrial Research Organisation (CSIRO) titled "Australian National Outlook 2019". The report identified the following six major national challenges to be tackled by Australians to achieve a more prosperous future:

1 The rise of Asia is shifting the geopolitical and economic landscape, placing Australia in a vulnerable position unless it adapts.

2 With technological change and Australia's inability to reverse its recent decline in educational performance, the future workforce could be ill-prepared for future jobs.

3   Climate change and environmental challenges pose a significant economic, environmental and social threat.

4   Our population is growing and ageing, putting pressure on cities, infrastructure and public services.

5   Trust in institutions has declined. Unless this can be restored, Australia will find it difficult to create the long-term solutions required to address challenges.

6   Social cohesion measures have declined, with many Australians feeling left behind.

While the report makes a point that it's possible to leverage Australia's many positive qualities and strengths as a nation, and identifies the required targets to maintain economic growth, it lacks the much needed "how" information to achieve it. Perhaps that will be the next phase, and hopefully we'll be pleasantly surprised. However, a review of the current state of leadership, how leaders get to the top, and the systemic factors that have led to our current state of affairs in Australia and globally point to a less-than-ideal landscape.

A REORDERING OF TRUST

The past few years have seen colossal transformation in the workplace which – according to the "2018 Deloitte Global Human Capital Trends" report – has been driven by technological, social, economic and political changes. A major outcome from such complex factors is people's loss of trust in their political and social institutions, as we just discussed. Another outcome is the expectation that business leaders will fill the gap. But their research points to a different narrative.

Traditional power elite figures, such as CEOs and heads of state, have been discredited, largely thanks to the growth of social media platforms. These communication channels have shifted people's trust from a top-down orientation to a horizontal one in favour of peers and influencers. According to Edelman's global survey results, we are now seeing a reordering of trust to more local sources, with "My Employer" emerging as the most trusted entity. This is believed to be because the relationships that are closest to us feel more controllable.

Employee expectations that prospective employers will join them in taking action on societal issues (67%) are nearly as high as their expectations of personal empowerment (74%) and job opportunity (80%). Sixty percent look to their employer as a trustworthy source of information on contentious social problems and on important topics like the economy (72%) and technology (58%). Interestingly, the general population sees business as being able to make money *and* improve society (73%).

This shift in employee expectations opens up enormous opportunities for employers to help rebuild societal trust. Edelman offers a path for businesses to follow to help restore trust, which – according to them, and in the CSIRO's report previously mentioned – is the greatest moral challenge of our era. Their call to action is for the workplace to build a better future for all, with the following suggested actions:

1   **Lead change.** Establish an audacious goal that attracts socially minded employees and make it a core business objective. Companies must first address employees' real fears about the threat of automation to their livelihoods and provide them with retraining if needed.

2   **Empower employees.** Go direct to employees with fact-based information about the issues of the day, supplementing the mainstream media. Give employees a voice on company channels; trust in company-owned media rose by seven points in 2019.

3   **Start locally.** Care for the communities where you operate, especially if you're a multinational. Be part of the solution on education, inequality and infrastructure. Enable your employees to volunteer and give back locally.

4   **CEO leadership.** CEOs must speak up directly on social issues such as diversity, inclusion and refugees. They must walk the talk and demonstrate their personal commitment inside and outside the company. Seventy-six percent of people surveyed expect CEOs to take a stand on challenging issues.

There is substantial merit in Edelman's proposed employee–employer contract, aptly named "Trust at Work". To be sure, a fundamental rebalancing of the employee–employer relationship is needed, shifting from top-down control to one that emphasises employee empowerment. This is because in a full-employment economy, an employee has more freedom to choose the kind of workplace they're coming to expect, one where values and the power to make change are a given.

These are not new concepts. They are reminiscent of Stephen Covey's call for an empowered workforce and empowering leaders in his seminal 1989 book *The 7 Habits of Highly Effective People: Restoring the character ethic*, and in his 2004 book *The 8th Habit: From effectiveness to greatness*. Many organisations were transformed by leaders willing to embrace the principles taught in these books (I had the good fortune of working for one in the 1990s). On the other hand, we've witnessed three decades of subsequent "leadership trust erosion". This leads to the question, *what type of leaders have been making it to the top? And why?*

WHO GETS TO BE CEO?

If we were to do a Google search on the path to becoming a board member or CEO, we'd find thousands – if not millions – of results. We'd probably find the same number of results when searching for the characteristics of successful CEOs or board members.

Leadership research is relatively new, receiving attention from scholars only in the past 30 years, whereas the physical and natural sciences have been formally researched since the 18th century. Leadership research and concepts will be discussed further on. However, understanding who is making it to the boardroom and C-suite of large organisations provides an insightful backdrop to future discussions.

Harvard Business Review perspective

Usually regarded as the bastion of business knowledge (at least within business schools in the US and Australia), the *Harvard*

Business Review (*HBR*) publishes a yearly article titled "The Best-Performing CEOs in the World". Reviewing their lists of the world's top hundred CEOs from 2014 to 2018 reveals some interesting information:

- the top 10 CEOs were completely different after four years
- there were no women in the top 20
- their criteria was based on the companies appearing in the S&P Global 1200, which is an index that comprises 70% of the world's stock market capitalisation and includes companies in North America, Europe, Asia, Latin America and Australia
- there was only one Australian CEO, ranked in 96th place in 2018.

In their 2018 report, *HBR* mentioned that to produce their top 100 list they had consulted with two providers of environmental, social and governance (ESG) research and analytics, which is a step in the right direction. However, it seems that the overarching criteria for a best-performing CEO according to *HBR* is mostly based on financial performance in the stock market.

Other interesting facts about CEOs from the same *HBR* report include:

- 20% lead companies based outside their countries of birth
- 32% have an MBA
- 34% have an engineering degree
- 16% became CEO at age 44 and have been in office 16 years
- 3% are women.

LinkedIn and Investopedia perspectives

Other interesting data comes from LinkedIn (the leading social media platform for business) and Investopedia (the Wikipedia of financial education). In 2018, LinkedIn looked at 12,000 profiles with the job title CEO from 20 countries and found that today's

CEOs are not a diverse bunch, with less than 5% of Fortune 500 CEOs being women and only three Fortune 500 CEOs being black. They found that computer science was the most popular field of study by a large margin. That was followed by economics and business. Thirty-three percent had MBAs. The first job held by most CEOs was in a business development function, by a large margin. Then came sales, engineering, IT and consulting. Twenty percent were promoted internally and 80% externally. It's important to note that the methodology used for their analysis is far from robust, and these results ought to be viewed as such. Nonetheless, these are still interesting figures.

In April 2019, Investopedia looked at the United States' top CEOs and their academic qualifications, and although it wouldn't be correct to compare their findings with *HBR*'s and LinkedIn's (they each followed their own criteria), it's still worthwhile noting a few findings. While the others do not say it explicitly, their lists reflect a point made by Investopedia from the outset: while there are a few exceptions, like Bill Gates, Steve Jobs and Mark Zuckerberg, a formal education is standard for the top tier of the business world. The path to becoming a Fortune 500 CEO in the US includes more than one university degree. In addition, they found that it most often includes degrees from the country's top universities.

The leadership literature perspective

To shine more light on career pathways of today's CEOs, *The Oxford Review* conducted a special research review of the leadership literature in 2018. These research findings will be referred to throughout the book, especially in relation to pathways and potential obstacles encountered by STEMM leaders in reaching senior leadership roles.

Here is a summary of *The Oxford Review*'s findings:

- The more selective the university the leader attended – for example, Oxford, Harvard, Cambridge – the more likely they are to be appointed as a CEO. However, their education level has not been found to be linked to the organisation's performance.

- The type of degree the CEO holds has an impact on the company's R&D funding. Those with STEMM qualifications will spend more on R&D.

- CEOs with MBAs tend to be more aggressive and have higher levels of capital expenditures, build more debt, and pay smaller dividends.

- CEOs with higher degrees in management subjects tend to have more fixed ideas about how they believe decisions should be made and how strategy should be developed and prefer administrative complexity.

- Entrepreneurial CEOs (that is, those with innovative and risk-taking propensity, positive awareness, and perseverance and endurance) tend to have a higher academic level in technical and engineering subjects and have R&D and marketing knowledge and greater breadth of experience.

- CEOs from another culture (not the home culture of the business) or CEOs who've lived abroad tend to foster greater internationalism in their firms.

- Organisations with materialistic CEOs (psychology defines "materialism" as a way of life where an individual displays an attachment to worldly possessions and material needs and desires) tend to be less adept at managing risk, foster more aggressive trading and marketing behaviours, and are more likely to engage in "creative" regulatory interpretations, especially during a financial crisis, putting the organisation at further risk.

- Overconfident CEOs and CFOs are significantly more likely to engage in tax avoidance.

- CEOs and board members with a higher educational background and greater expertise and experience tend to contribute to higher levels of organisational performance and lower risk.

- CEOs who engage in uncertainty avoidance behaviours (denial and avoidance in uncertain situations and the creation of false certainty) tend to promote lower levels of corporate social responsibility and performance and have higher levels of narcissistic characteristics.

- CEOs with high levels of cognitive flexibility (the ability to take in peripheral and diverse information and change decisions and ways of thinking) and an orientation towards uncertainty and complexity were found to create more adaptive organisations.

OTHER INFLUENTIAL LEADERSHIP ROLES: HEADS OF STATE

It's difficult, and beyond the scope of this book, to establish a correlation between the qualifications and background of national heads of state. However, a review of various reputable online sources points to the fact that, at least here in Australia, and in one of our closest allies, the United States of America, there are no specific requirements for the job. Notwithstanding the lack of selection rigour, there are several obvious patterns.

The Prime Minister of Australia is the head of the government of Australia. Most of Australia's prime ministers have been Australian-born, middle-aged, tertiary-educated men with experience in law or politics, representing electorates in either Victoria or New South Wales. Only one woman has served as Prime Minister. The average age is 52 years, which reflects the age profile of Australian parliamentarians more generally (51 years). Three-quarters of Australia's 29 prime ministers (22) were born in Australia. Of those born overseas, all but one came from the United Kingdom (England, Scotland or Wales). The only non-British overseas-born Prime Minister was Chris Watson, who was born in Chile and raised in New Zealand. Of those born in Australia, the majority were born in either Victoria (nine) or New South Wales (eight).

Our PM at the time of writing, Scott Morrison, has a Bachelor of Science in applied economic geography from the University of New South Wales (UNSW), although he never worked in science. His CV has been widely described as that of a career politician who leveraged his extraordinary networking and marketing skills within the right circles to reach the top.

In the United States Constitution there are three requirements a person must fulfil to be eligible for the presidency. Candidates must be at least 35 years old, have 14 years of residency within the US, and be a natural-born US citizen. There are no education requirements to become president, as evidenced by the educational history of the 43 men who have become president so far. Nine presidents never went to college, Harry Truman being the most recent. Many of them – 25, including Barack Obama – were lawyers, though not all 25 completed law school. Four were professional soldiers. One president, William Harrison, attended medical school.

The current President is Donald Trump. He has a background as a real estate businessman and has a degree in Economics from the Wharton School of Finance, University of Pennsylvania. Trump was also a reality TV personality, gaining international fame with his show *The Apprentice*.

WHO ARE WE PROMOTING AND ELECTING? AND WHY?

In 2013, the then Vice President of R&D of Hogan Assessments and organisational psychologist Tomas Chamorro-Premuzic wrote the *Harvard Business Review* article "Why Do So Many Incompetent Men Become Leaders?". The article was published just a few months after Sheryl Sandberg took the world by storm with her 2013 book *Lean In*, telling women to forge ahead in their careers while juggling the balancing act with parenthood. Perhaps the timing of Chamorro-Premuzic's article was coincidental, or perhaps not. However, we now had a high-profile psychologist finally "calling it out". I had the good fortune of watching him present at a Hogan event in Sydney in 2016 and was enthralled by his depth and breadth of knowledge about psychometric testing (as a Hogan-certified assessment practitioner, and I brought along one of my clients from UNSW Business School).

As an expert in personality science and assessments, Chamorro-Premuzic spoke energetically about the current situation concerning leadership. He has gone on to publish books on the topic, which cover certain common tenets. In general, people commonly misinterpret displays of confidence as a sign of competence. Regardless of nationality, manifestations of charisma or charm are commonly mistaken for leadership potential, and unfortunately these occur much more frequently in men than in women. This helps explain the ridiculously low numbers of female CEOs and heads of state globally. And when women do try to display the same types of behaviours, they are subjected to a whole different set of biases. The world often associates hubris with competence. Chamorro-Premuzic went on to assert that, according to research, pretty much anywhere in the world men tend to think that they are much smarter than women.

The same psychological characteristics that enable certain male managers to rise to the top of the corporate or political ladder are also responsible for their downfall. That is, what got them to the top is not what it takes to do that job well. As a result, too many incompetent people are promoted to management jobs over more competent people. As we'll see in later chapters, the characteristics needed for effective leadership are not the ones we've historically promoted; for example, narcissism (egotistical, self-absorbed), psychopathy (no empathy, no remorse), or histrionic (theatrical, dramatic).

Not surprisingly, the mythical image of a leader embodies many of the characteristics found in personality disorders, as discussed by the world-renowned leadership expert and psychiatrist Manfred Kets de Vries in his 2014 book *Mindful Leadership Coaching: Journeys into the interior*. And as we've already observed locally and internationally – whether in politics or business – the question, "*Why do so many incompetent men become leaders?*", sadly, remains relevant.

Political and business leaders are largely failing

So far, the evidence continues to indicate that political and business leaders are largely failing their followers and subordinates, and most of us continue to experience leadership in a negative way. More bosses are contributing to burnout, anxiety, boredom and loss of productivity than driving high-performing teams and organisations.

As solutions-focused, evidence-based coaching psychology practitioners, executive coaches are certain that much can be done to create improvements. Chamorro-Premuzic quips that in "an ideal world", leaders would follow science-based practices and prioritise engaging with their employees, providing them with meaning and purpose. He also equates positive leadership behaviours (which will be discussed further on) with being feminine. On the other hand, he describes hubris as being masculine. Perhaps therein lies some of the problem, and potential solutions.

Society has not advanced enough to view femininity as a positive attribute in leadership circles. And as we'll soon see, positive feminine attributes include empathy and emotional intelligence (EI), which tests have shown women to be superior in when compared to men. Perhaps men are unconsciously rejecting the call to be more empathetic and to develop more EI. This is a real issue, and one that's been raised by some of my male clients, especially those in roles and organisational cultures that place a premium on the masculine interpretation of confidence; for example, being loud, almost aggressive, perhaps mischievous, pushy, unyielding, with a win–lose mindset.

Indeed, one of our clients, Rob, a software engineer in his early thirties who quickly worked his way up to become Sales Director within one of the largest IT multinationals, found himself behaving like two different individuals: one who seemed to be emotionally intelligent and empathetic with some colleagues, and the other who was bullish and "rough" with other colleagues. In our coaching discussions, he admitted to wanting to fit in with the rest of the senior leadership team and his CEO, who were significantly older than him and behaved, quite often, in a very "rough" manner (which he described as the "Aussie macho" culture). As a young and progressive senior leader working in a masculine organisational culture, he desperately wanted to assert himself with his peers and boss.

This dichotomy spilled over at home and was even highlighted by members of his family.

Interestingly, Rob completed an MBA at an elite university in the late 2000s. In those days, MBAs were skewed to emphasise economic, financial and marketing success (some might say, at almost any cost). Course curricula contained limited evidence-based leadership practices for sustainable organisational performance and employee wellbeing, let alone crucial leadership capabilities and skills like emotional intelligence and principled decision making. Then, perhaps, we shouldn't be surprised to encounter leaders behaving unethically and with little integrity (here, integrity is defined as doing what they say they'll do, even when nobody is watching).

If this is what is taught in elite business schools as recently as 2011, the events leading up to and following the Australian Banking Royal Commission shouldn't come as a shock. As *The Guardian* eloquently showed in their April 2018 article "A Recent History of Australia's Banking Scandals", outrageous moments in the banking and financial sector in the nine years since the Global Financial Crisis have left us questioning the integrity of our leaders. Hence, the lack of trust mentioned earlier.

We need to redefine leadership and describe the right attributes according to the evidence. We must ask ourselves: what are the qualities and characteristics a person must possess to lead others to achieve common goals, have meaning and enjoy wellbeing in the workplace and in life? Fortunately, there's a wealth of evidence-based research to help us answer that question. It has been the effort of countless researchers around the world, applying scientific approaches, proving once more the value of rational thinking. Fields like Positive Organisational Psychology, Coaching Psychology, Neuroscience and Leadership Theory and Practice are converging with Medical Sciences to help us demystify leadership, not as an artform or something people are born with, but rather as a process that can be applied and learned.

CHAPTER 2

LEADERSHIP DEFINED
AND WHY IT MATTERS

*"If your actions inspire others to dream more, learn more,
do more and become more, you are a leader."*

John Quincy Adams

There are many ways to finish the sentence, "Leadership is ... .". There are possibly as many different definitions of leadership as there are people who have tried to define it. Although each of us intuitively knows what we mean by the word "leadership", it can have different meanings for different people. One thing is certain though: leadership is a highly sought after and highly valued commodity. The public has become increasingly captivated by the idea of leadership, and people continue to ask themselves and others what makes a good leader. And yet, scholars and practitioners have attempted to define leadership for more than a century without a universal consensus.

For most of us, notions of good and bad leadership are derived from our own careers. In fact, a popular leadership coaching or workshop exercise is to ask people to think about the worst boss they ever had and write down their characteristics. This is then followed by identifying the best boss they ever had and listing their characteristics. Not surprisingly, in the 10 years I've been coaching, I've found that both good and bad characteristics are consistent across industries, age groups, genders, qualifications and organisations. The following table summarises the common characteristics individuals in our leadership workshops use to describe a good leader versus a bad one.

Figure 5: Qualities of a bad vs a good leader (client feedback)

Bad leader	Good leader
Poor communicator	Communicates effectively
Over-confident, arrogant	Humble
Controlling, tells you what to do	Empowers, doesn't control
Doesn't ask, or listen	Asks and listens actively. Coaches others. Puts the needs of the team first
Takes all credit	Puts the limelight on the team
Doesn't share information	Shares information and helps makes sense of hard decisions
Doesn't give positive feedback	Generous with time, praise and feedback
Shows preferential treatment	Brings out the best in everyone, not just a few
Doesn't set a vision	Sets a compelling vision
Leads through fear	Inspires, does what they say they'll do; i.e. integrity
Doesn't walk the talk	Leads by example and side by side with the team
Never says "please" or "thanks"	Respectful, civil, empathetic

Our leaders' behaviours have a significant influence throughout our lives, informing our own perceptions of the world and decision-making. This is because emotions and behaviours are contagious, as shown by extensive research in psychology and neuroscience. If you have a friend who is happy, the probability that you will be happier rises by 25%. If you have overweight friends, you're more likely to be overweight yourself. If you quit smoking, your friends are more likely to quit. This is known as "social contagion", and it also applies to leaders.

Most readers would already know that good leadership creates engaged employees and that leaders influence a variety of outcomes such as staff turnover, customer satisfaction, productivity, revenue, innovation and growth. However, what most fail to understand is that a good leader can make other people around them good leaders. Indeed, studies show that there are certain leadership behaviours that are very contagious. Here are the most contagious out of 51 behaviours tested (in order of most to least contagious) according to a study reported in a 2016 *HBR* article titled "The Trickle-Down Effect of Good (and Bad) Leadership":[3]

- Developing self and others
- Technical skills
- Strategy skills
- Consideration and Cooperation
- Integrity and Honesty
- Global perspective
- Decisiveness
- Results focus.

Sadly, the same applies to negative and counterproductive behaviours, as described in a recent study in the *Journal of Applied Psychology* titled "Catching Rudeness is like Catching a Cold: The contagion effects of low-intensity negative behaviours". Rudeness and incivility have been shown to decrease productivity at work, not to mention the other side effects like a negative work climate due to fear and lack of psychological safety.

If we take a moment to think about the occasional things we do poorly and the bad habits we can't seem to change, perhaps

---

3   Zenger, J., & Folkman, J. (2016). "The Trickle-Down Effect of Good (and Bad) Leadership". *Harvard Business Review.* Retrieved from https://hbr.org/2016/01/the-trickle-down-effect-of-good-and-bad-leadership.

considering this research might motivate us to change. After all, there's a high probability we're being mimicked by others like our peers, our direct reports, our partner or spouse, and our children. Luckily, there is a great deal we can do to protect our family and our team from our blunders and unfortunate habits. We *can* change.

Too many new clients come to us wondering whether they are making an impact. Being unable to see our impact on others is quite normal since the influence may be subtle and may take many years before it becomes evident. Make no mistake though, our behaviour as a leader has an impact much greater than we might have suspected. Leaders cast a long shadow and do make a difference.

How can we do better?

RE-CONCEPTUALISING LEADERSHIP

In the past 60 years, as many as 65 different classification systems have been developed to define the dimensions of leadership. Some definitions view leadership as the focus of group processes. From this perspective, the leader is at the centre of group change and activity, and embodies the will of the group. Another set of definitions conceptualises leadership from a personality perspective, suggesting that leadership is a combination of special traits and characteristics that some individuals possess. Other approaches to leadership define it as an act or a behaviour – the things leaders do to bring about change.

Others define leadership in terms of the power relationship that exists between leaders and followers. From this viewpoint, leaders have power that they wield to effect change in others. In addition, others view leadership as a transformational process that moves followers to accomplish more than is usually expected of them. Finally, some scholars address leadership from a skills perspective, stressing the capabilities – knowledge and skills – that make effective leadership possible.

Leadership science

The reason there are myriad definitions for leadership is because there is no universally agreed scientific approach to leadership; that is, a theory of knowledge, especially with regard to its methods, validity, and scope, and the distinction between justified belief and opinion.[4] In the field of leadership studies there are many competing approaches, such as organisational psychology, economics, statistics, education and management studies.

The special characteristic of scientific inquiry, as STEMM professionals know, is that science deals with identifying causation. Therefore, leadership as a science would mean identifying the causal factors for leadership. That would certainly challenge previous theories that simply describe the traits, qualities and behaviours of a leader, because simply describing things in this fashion is not scientific inquiry – it lacks an accompanying theory of causation. For example, people might have accurately observed that before every battle led by Alexander the Great, a black crow was seen holding a pine branch in its beak. It might have therefore seemed like a good idea to have a priest look for a crow with a pine branch in its beak before committing the army to battle. This is how mysticism becomes epistemology. Mysticism lacks any attempt at explaining cause and effect. A modern illustration of this is the popular notion that great leaders, particularly in business, were often academic underachievers. This is an inductive claim, not a scientific one where the element of causation has been identified.

An issue currently affecting leadership research – in the modern era of social sciences and econometrics – is when hypotheses arise from empirical observation in laboratories or by data analysis. We must be mindful about viewing leadership science as the empirical observation of a trait in leaders and after statistical analysis using it to specify what traits predict leadership behaviour. Again, this doesn't consider causation, and so we would be wise not to place high weighting on these predictions when selecting and developing leaders. Leadership science is therefore not necessarily about

---

4   Bret N. Bogenschneider, School of Law Senior Lecturer at the University of Surrey.

mathematical proof, but rather it is about a systemisation of theories which make the identification of cause and effect possible.

A general theory of leadership would therefore be: an object person (a leader) causes the subject group (group members) to proceed with a project (a goal) despite adversity (individual and systemic obstacles) with decisive effect (the result).

Approximately 30 leadership definitions (collated by the University of Warwick in 2012) were assessed against this general theory of leadership. A clear, inclusive and relatable approach to defining leadership was given by Peter G. Northouse, Professor Emeritus of Communication at Western Michigan University, in the seventh edition of the 2016 book *Leadership: Theory and practice*, where he identified the key components central to the phenomenon:

- leadership is a process
- leadership involves influence
- leadership occurs in groups
- leadership involves common goals.

By bringing together the above components, the following scientific definition of leadership is used in this book (and all our leadership development programs):

Leadership is a process whereby an individual influences a group of individuals to achieve a common goal.[5]

Defining leadership as a *process* means that it's not a trait or a characteristic that resides in the leader, supporting the notion that leaders are not necessarily born but made. "Process" implies that a leader affects and is affected by followers. It emphasises that leadership is not a linear, one-way event, but rather an interactive and iterative event where continuous learning can occur. Indeed, when leadership is defined in this manner, it becomes available to everyone,

---

5   Northouse, P. (2016). *Leadership: Theory and Practice* (7th ed.). Los Angeles: SAGE.

which is great news. Nor is it restricted to the formally designated leader in a group.

As implied by the visual conceptualisation of leadership as a process in figure 6, this approach demystifies leadership as an artform, which often confuses STEMM professionals when contemplating a career in management and, later, in leadership. It renders leadership a reachable field, one that can be studied, observed, improved upon … much like the scientific method.

Figure 6: Trait definition versus Process definition of leadership

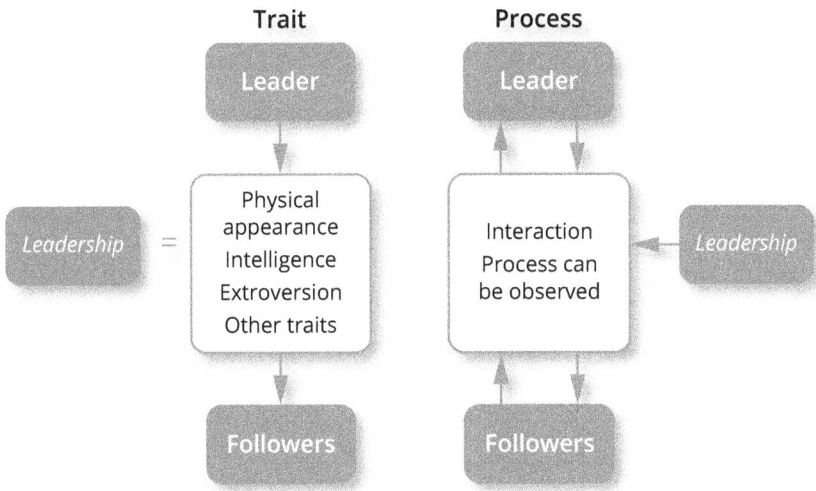

IS CURRENT LEADERSHIP DEVELOPMENT WORKING?

We live in a new reality. People are looking for "heroic leaders" who inspire at every level. Business leaders today are public figures, almost celebrities. People like Richard Branson, Bill Gates, Sheryl Sandberg, Jeff Bezos, Arianna Huffington, Mark Cuban and Warren Buffet are household names. And yet, original leadership research specific to the 21st century remains thin, as observed by Barbara Kellerman, executive director of the Centre for Public Leadership at the Harvard Kennedy School. She has been vocal for many years about the leadership industry, calling for a rigorous rethinking of leadership theory and practice by adopting a more systemic perspective. She has said: "Our efforts at leadership education, training and development are miserably inadequate ... Getting leadership right – or, at least, a lot more right than what we do now – would go a long way toward addressing the problem."

It's been estimated that approximately 50% of senior leaders fail or have left within 18 months of a new appointment. And those making the transition from operational management to leadership roles will grapple with the shift from "the known", within defined parameters, to "the unknown".

Other challenges faced by leaders include:

- Problem-solving fatigue and inertia

- Shifting employment positions, roles and processes

- Changing or unfamiliar cultures, business policies, practices and processes

- Conflicting business philosophies, perspectives and options

- Information overload.

According to a 2019 *Harvard Business Review* article titled "The Future of Leadership Development", the need for leadership development has never been more urgent. Companies of all kinds realise that to survive in today's *volatile, uncertain, complex* and *ambiguous* environment (known as VUCA), they need leadership skills and organisational capabilities *different from those that helped them succeed in the past*. The article goes on to recognise that leadership development should not be restricted to the few who are in, or close to, the C-suite. And yet, we find that organisations invest their leadership development budget mostly on their senior leaders. Indeed, 90% of our leadership coaching clients are in the C-suite or one level below. We need to start developing leaders much earlier than that if we are to embed the necessary mindset of continuous learning, adaptability and growth, which is scientifically supported by adult development theory. (This will be discussed further on.)

With the proliferation of collaborative problem-solving platforms and digital "adhocracies" that emphasise individual initiative, employees across the board are increasingly expected to make decisions that align with corporate strategy and culture. It is therefore important that they are equipped with the relevant technical, relational, and communication skills. And yet, organisations that collectively spend billions of dollars annually to train current and future executives continue to be frustrated with the results. Doing an MBA and/or other types of classroom-based courses is no longer enough – and as implied by the data, perhaps it never was.

The three main reasons for the disjointed state of leadership development

As with all challenges, it's best to look at the potential root causes; they help us unearth what does work and what needs improvement. Here are the three main reasons for the disjointed state of leadership development:[6]

1 **There is a gap in motivations.** Organisations invest in executive development for their own long-term benefit, however individuals participate to enhance their skills and advance their careers, and may end up leaving the employer that has paid for their training. This is a catch-22 situation, because if companies *aren't* committed to developing their staff, people will have greater reason to move on. Companies need to look at ways of providing meaningful career paths and an engaging environment to retain their employees.

2 **There is a gap between the skills that executive development programs build and those that firms require today.** This is particularly relevant in the areas of interpersonal and metacognitive skills (awareness of your own thought processes), which have become extraordinarily essential in today's hyper-connected and networked world where the ability to collaborate with others has a direct impact on an organisation's success. Traditional providers of MBAs and executive education programs bring deep expertise

---

6   Moldoveaunu, M., & Narayandas, D. (2019). "The Future of Leadership Development (pp. 40–48)." *Harvard Business Review*, March–April 2019.

in disciplines such as economics, accounting, marketing, management, finance and strategy. Clearly, these are essential for functional managers with deep technical expertise in one discipline. They seek general management knowledge, and for this, MBAs serve an excellent purpose. However, leaders are now in dire need of more sophisticated interpersonal, communication, and relational skills. Unfortunately, these have been historically underplayed by business schools.

3 **There is a skills transfer gap.** Few executives seem to apply what they learn in the classroom to their jobs. In addition, the further removed the context of learning is from the context of application, the larger the gap becomes. Basically, traditional executive education seems to be too episodic, exclusive and expensive to achieve leaders' goals of "lifelong learning". The good news is, technology has opened new possibilities for everyone, not just the elite few.

THE RISE OF PERSONALISED LEARNING

The focus for many organisations this past decade has been on becoming adaptive and agile, according to Deloitte's 2016 "Global Human Capital Trends". As they strive to become more agile and customer-focused, organisations are shifting their structures toward interconnected, flexible teams. Agility and adaptiveness are also expected in learning and development (L&D) initiatives. Employees at all levels expect dynamic, self-directed, continuous learning opportunities from their employers. Not surprisingly, the top-rated trend for 2019, in Deloitte's "Global Human Capital Trends", is the need to improve L&D.

With the advent and rapid development of cloud technology, personalised learning ecosystems are now possible. Examples include:

- Massive open online courses (or MOOCs) and platforms such as Coursera

- Corporate L&D systems from LinkedIn Learning and other skills mastery courses on interactive platforms
- On-demand leadership development from the likes of McKinsey Academy
- Talent management platforms which connect learning needs and learner outcomes to recruitment, retention and promotion decisions.

However, as the 2019 survey by Deloitte showed, expectations and reality are two very different things. Digital HR is still aspirational, and providers have some major work ahead before it's truly meeting the expectations of employers, employees and HR leaders. There is no question though, we've only just been scratching the surface in terms of personalised learning technology and human experience, with neuroscience, medicine and coaching psychology now playing pivotal roles in leadership development and overall employee flourishing in the 21st century.

Technology and scientific advancements were indeed at the centre of the 2018 Coaching in Leadership and Healthcare Conference in Boston, organised by the Harvard Medical School Institute of Coaching (where I am a Fellow member). This is where STEMM research meets coaching psychology theory and gets translated into practical applications for leadership development and wellbeing programs for positive outcomes at all levels: individual, team and organisational. Due to its link to the Harvard Medical School, the Institute of Coaching is at the forefront of scientific, evidence-based methodology.[7]

THE FOURTH INDUSTRIAL REVOLUTION

This is a great segue to introduce the relevance and impact that executive coaching has had in the workplace over the past 10 years,

---

[7]   Their Scientific Advisory Board consists of internationally respected researchers and thought leaders in various fields. One of its most prominent members is Professor Anthony Grant – the Director of the Coaching Psychology Unit in the Department of Psychology at The University of Sydney, and a pioneer in coaching psychology. He set up the world's first tertiary degree in coaching psychology, and I had the privilege of studying with him.

which will be discussed in greater detail in later chapters. For now, the takeaway messages from this chapter would have to include that we're about to witness seismic shifts in the future of learning and development (L&D) and of leadership development. As echoed by Moldoveaunu & Narayandas in the 2019 *HBR* article, employee development will likely have the following characteristics, which individual and team coaching programs can provide:

1 **Learning will be personalised.** Leaders and employees will pursue the programs and practices that are right for them, at their own pace, at any time that works for them.

2 **Learning will be socialised.** Learning often happens best when learners collaborate and help one another. Knowledge is social in nature and platforms will provide groups with opportunities to solve problems together.

3 **Learning will be contextualised.** Most executives value the opportunity to get professional development on the job, in ways that are relevant to their work environment. As I always say, one size does not fit all.

4 **Learning outcomes will be transparently tracked, and in some cases, authenticated.** Just because learning programs are completed and stored in the cloud doesn't imply devaluing credentials like diplomas and degrees. On the contrary; it will drive a new era of capabilities-based certification and mastery. Mastery depends on practice and feedback, and the cloud is perfect for the management of interactive activities that emphasise mutual feedback and allows executives to learn on the job while doing the work they always do.

5 **Learning will be evidence-based.** This was not mentioned in the *HBR* article, however it must be included. There are thousands of service providers out there, and thousands more are likely to pop up in the next few years. HR will soon be required to ensure that L&D methods and tools have been scientifically shown to work. This must be part of the purchasing criteria for decision-makers in organisations, not just cost and a clever marketing pitch.

As the evidence shows, executive coaching is one of the most effective approaches to changing mindsets and hence changing behaviours to yield tangible and intangible benefits. Whether coaching works or not is no longer in question, and it is now a key component of most L&D strategies employed by HR leaders globally.

However, coaching opportunities must also be made available to STEMM professionals as early as possible in their careers. These programs must be specifically designed to cater for their needs, and for the unique systemic obstacles within their organisations. In doing so, more STEMM professionals will develop the essential interpersonal and strategic-thinking skills to participate in crucial leadership discussions where the final decision can have significant consequences to the business, the organisation and to society. Having more STEMM leaders at the decision-making table – in the boardroom as it were – may be the missing ingredient in any organisational transformation strategy and execution plan.

To understand why this needs to be made a priority, we must first understand what makes the 21st century so different and the problems we now face.

CHAPTER 3

THE FOURTH INDUSTRIAL REVOLUTION AND STEMM LEADERSHIP

"I do not share your view that the scientist should observe silence in political matters."

Albert Einstein

Klaus Schwab, founder and executive chairman of the Geneva-based World Economic Forum (WEF), published a book in 2016 titled *The Fourth Industrial Revolution* and spread the term at the Davos meeting that year. The Davos meeting convenes world leaders to discuss global, regional and industry agendas at the beginning of each year. Back then, Schwab predicted that we were standing on the brink of a technological revolution that would radically alter the way we live, work and relate to one another.

In its scale, scope and complexity, the transformation will be unlike anything humankind has experienced before. We're already observing a series of social, political, cultural and economic upheavals unfolding in the 21st century. The Global Financial Crisis of 2007–08 caused social and financial devastation across the globe. And more recently, Brexit and the Trump administration are causing chaos in previously stable and reliable democracies. And there are myriad social, financial and political tensions flaring up in all corners of the world.

Building on the widespread availability of digital technologies that were the result of the Third Industrial Revolution (or the Digital Revolution), the Fourth Industrial Revolution will be driven largely by the convergence of digital, biological and physical innovations.

As figure 7 illustrates, the First Industrial Revolution is remembered for its steam-powered factories, the Second for the application of science to mass production, and the Third for digitisation. The Fourth Industrial Revolution is about technologies such as artificial intelligence (AI), genome editing, augmented reality, 3D printing, and robotics, which are rapidly changing the way humans create, exchange and distribute value. Our ability to edit the building blocks of life and transform our society have been expanded by:

- Low-cost gene sequencing and techniques such as CRISPR

- Artificial intelligence, which is augmenting processes and skills in every industry

- Neurotechnology that is making unprecedented strides in how we can use the human brain

- Automation that is disrupting century-old transport and manufacturing paradigms

- Technologies such as blockchain and smart materials, which are redefining the boundaries between the digital and physical worlds.

These changes will profoundly transform businesses, societies, economies, institutions, industries and individuals, as did previous industrial revolutions. However, the sense that new technologies are being developed and implemented at an increasingly faster pace is having an impact on human identities, communities and political structures.

The three previous revolutions have had both positive and negative impacts on different stakeholders. Nations have become wealthier and technologies have helped pull entire societies out of poverty, as discussed in chapter 1 under the heading "Is the world getting better or worse?" However, our inability to distribute benefits fairly or anticipate externalities (in economics, an externality is the cost or benefit that affects a party who did not choose to incur that cost or benefit) has resulted in global challenges.

Figure 7: The Four Industrial Revolutions, from the 18th to the 21st century

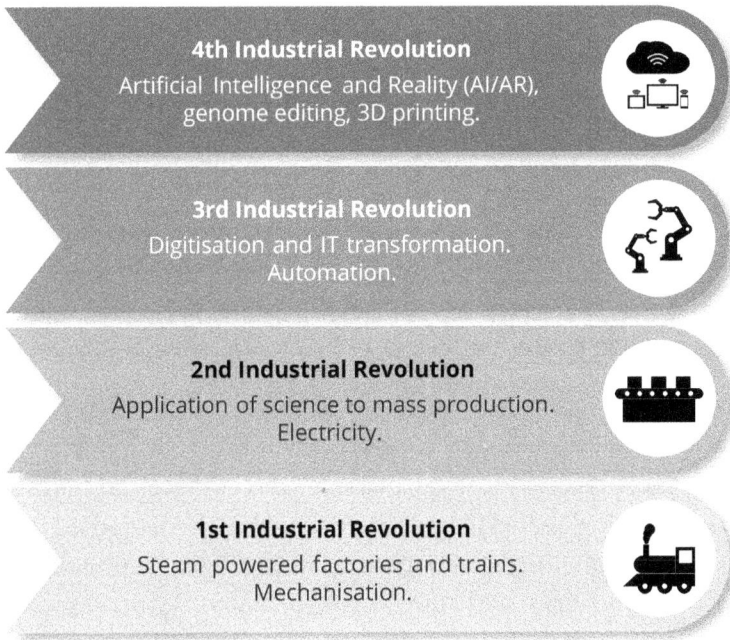

4th Industrial Revolution
Artificial Intelligence and Reality (AI/AR), genome editing, 3D printing.

3rd Industrial Revolution
Digitisation and IT transformation. Automation.

2nd Industrial Revolution
Application of science to mass production. Electricity.

1st Industrial Revolution
Steam powered factories and trains. Mechanisation.

Image: Shutterstock.com

We need to recognise these risks; whether it's cybersecurity threats, misinformation on a massive scale through digital media, potential unemployment, or increasing social and income inequality, we must take steps to align common human values with our technological progress and ensure the Fourth Industrial Revolution benefits all human beings first and foremost. With these fundamental transformations underway now, in 2019 and beyond we could proactively shape the Fourth Industrial Revolution to be both inclusive and human centred. In a way, given the enormous magnitude of the UN Global Goals, the Fourth Industrial Revolution may be an enabler for developing, diffusing, and governing technologies that will allow us to achieve the Global Goals.

We need to govern technologies in ways that foster a more empowering, collaborative and sustainable foundation for social and economic development built around values that support the common good, human dignity and human rights. To do this, we need to adapt and improve governance and leadership models, beginning with embracing diversity of thought within organisational leadership. We are already seeing significant changes in the types of jobs that are becoming more relevant, and those that are gradually disappearing in what has become commonly termed "The Future of Work".

NEW LEADERSHIP IS NEEDED FOR THE 21st CENTURY

As the Fourth Industrial Revolution unfolds, organisations are seeking to harness new and emerging technologies to reach higher levels of efficiency, expand into new markets, and compete on innovative products for a global consumer base increasingly composed of digital natives (individuals who grew up in the digital age, rather than acquiring familiarity with digital systems as an adult). Employers are therefore also seeking employees with new skills, as detailed in the report by the WEF: "The Future of Jobs Report 2018".

Business leaders across all industries and regions will be called upon to formulate a comprehensive workforce strategy ready to meet the challenges of this new era of accelerating change and innovation. Policy makers, educators, labour groups and individuals have much to gain from a deeper understanding of this new workforce landscape. As figure 8 shows, there will be a continued fall in demand for manual skills and physical abilities. In 2022, while proficiency in new technologies will be crucial, so will "human" skills such as creativity, originality, initiative, critical thinking, persuasion, negotiation, attention to detail, resilience, flexibility and complex problem-solving skills. Emotional intelligence, leadership, social influence and service orientation will also remain in high demand.

Figure 8: Comparison of skills demand in 2018 versus 2022

Skills needed in 2018	Skills needed in 2022	Skills in less demand in 2022
Analytical thinking and innovation	Analytical thinking and innovation	Manual dexterity, endurance and precision
Complex problem-solving	Active learning and learning strategies	Memory, verbal, auditory and spatial abilities
Critical thinking and analysis	Creativity, originality and initiative	Management of financial, material resources
Active learning and learning	Technology design and programming	Technology, installation and maintenance
Creativity, originality and initiative	Critical thinking and analysis	Reading, writing, math and active listening
Attention to detail, trustworthiness	Complex problem-solving	Management of personnel
Emotional intelligence	Leadership and social influence	Quality control and safety awareness
Reasoning, problem-solving and ideation	Systems analysis and evaluation	Coordination and time management
Leadership and social influence	Reasoning, problem-solving and ideation	Visual, auditory and speech abilities
Coordination and time management	Systems analysis and evaluation	Technology use, monitoring and control

Adapted from: Future of Jobs Survey 2018, World Economic Forum

The above table shows the increasing need for employees with analytical, innovation, critical-thinking and complex problem-solving skills, which STEMM professionals possess. Driven by global attention to the socioeconomic impact of digital technology and AI, the understanding that innovation is the main driver of economic growth, and international education reforms to focus on STEM education, the Australian Government launched the $1.1 billion National Innovation and Science Agenda in 2015. Over $64 million has been allocated to fund early learning and school STEM initiatives under the Inspiring All Australians in Digital Literacy and STEM measure. Schools and universities around the country are promoting and investing in STEM education significantly more than before.

For a nuanced understanding of what this means in terms of leadership requirements to drive initiatives that will close the skills gap effectively, it's best to refer back to Deloitte's 2019 "Global Human Capital Trends" report, mentioned earlier. Their survey results, depicted in figures 9 and 10, suggest an urgent need for greater cognitive diversity among senior leadership circles. Cognitive diversity is defined as differences in thinking styles, knowledge, skills, values and beliefs among individuals within a group. The Fourth Industrial Revolution calls for new leadership mindsets, capabilities and skills if we are to flourish in the 21st century. This makes sense of course. As with the previous industrial revolutions, leaders evolve and emerge to meet the challenges of the times.

This isn't to suggest that all current leaders must step down and make way for new ones. It simply means that it's time to adapt to the new paradigms and consider opening the doors to potential or current leaders who might have been outside the narrow criteria currently in place. Maybe it's high time to think outside the "leadership box" and consider the outliers and original thinkers within the STEMM pool of professionals and leaders. But first, let's discuss the elephant in the room ...

Figure 9: A new context for 21st century leaders

Why do you think there is a difference in the unique requirements for 21st century leaders?	
Drivers	**% respondents who thought it applied**
New technologies	75%
Pace of change	66%
Changing demographics	57%
Changing customer expectations	53%

Note: only respondents who believed that 21st-century leaders had new and unique requirements answered this question.

Adapted from: Deloitte Global Capital Trends Survey 2019

Figure 10: New leadership skills for the 21st century

What do you believe are the unique requirements for 21st century leaders?	
Drivers	**% respondents who thought it applied**
Ability to lead through more complexity and ambiguity	81%
Ability to lead through influence	65%
Ability to manage on a remote basis	50%
Ability to manage a workforce with a combination of humans and machines	47%
Ability to lead more quickly	44%

Note: only respondents who believed that 21st century leaders faced new and unique requirements answered this question.

Adapted from: Deloitte Global Capital Trends Survey 2019

STEMM MATTERS

In 2017, the Australian Government issued the National Science Statement, meant to set a long-term approach to science and provide guidance for government investment, decision-making and clarity on strategic aims. All STEMM professionals, and even those who have not delved into any STEMM disciplines, would be wise to read Australia's National Science Statement because the government's vision is for an Australian society engaged in and enriched by science. It's therefore relevant to mention the four main objectives, shown in figure 11.

Figure 11: The Australian Government's objectives in National Science Statement

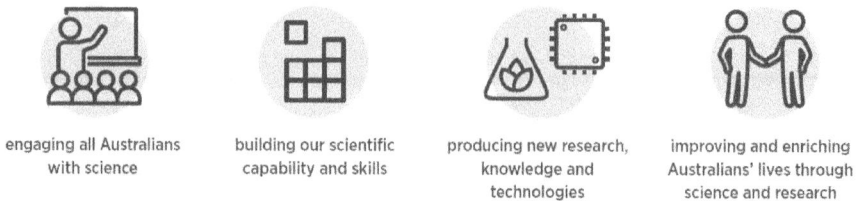

| engaging all Australians with science | building our scientific capability and skills | producing new research, knowledge and technologies | improving and enriching Australians' lives through science and research |

Source: Key points from Australian National Science Statement, science.gov.au/NSS

Notwithstanding the encouraging initiatives and steps taken to promote STEMM careers and knowledge in Australian society, there's a contrasting narrative in other national and global reports. As discussed in chapter 1, the WEF identified the world's failure to mitigate and adapt to climate change as one of the biggest threats of our time. Furthermore, Australia's own top scientific organisation, the CSIRO, also identified climate change as posing a significant economic, environmental and social threat globally and to Australia.

On the one hand, the government acknowledges the importance of scientific skills and education (to grow the economy?), and on the other it ignores the scientific community's call to take action to preserve the planet. We must therefore ask ourselves, *is our current leadership well equipped to make complex decisions of this nature?*

Before jumping to any conclusions, we should first develop a better understanding of the sources of scientific data and what the consensus of the global scientific community is on climate change.

The leadership of climate change

The Intergovernmental Panel on Climate Change (IPCC) was created by the United Nations Environment Programme (UN Environment) and the World Meteorological Organization (WMO) in 1988, and it has 195 member countries. Through its assessments, the IPCC determines the state of knowledge on climate change. The reports are drafted and reviewed in several stages, ensuring objectivity and transparency. The IPCC does not conduct its own research. IPCC reports are neutral; they are policy-relevant but not policy-prescriptive. The assessment reports, which can be easily accessed from their website, are meant to provide key input into the international negotiations to tackle climate change. However, the UN is not a regulatory body, therefore it's up to each sovereign nation to adopt policies in line with the IPCC findings and recommendations. The IPCC has nonetheless said: "Scientific evidence for warming of the climate system is unequivocal".

In the US, the National Aeronautics and Space Administration (NASA) has created a remarkable website titled Global Climate Change: Vital Signs of the Planet, which provides the general public, educators, students, and the media with easy-to-understand explanations and interactive media resources about climate change, as well as details about primary research sources. The website states: "Ninety-seven percent of climate scientists agree that climate warming trends over the past century are extremely likely due to human activities, and most of the leading scientific organizations worldwide have issued public statements endorsing this position." It's truly inspiring to see the efforts made by the scientific community to engage with the general community through many different types of programs.

In Australia, the CSIRO and the Bureau of Meteorology have been issuing the "State of the Climate" report for five years. The 2018 report discusses how Australia's weather and climate continues to change in response to a warming global climate. And, just like NASA's website (perhaps a tad less "flashy"), the State of the Climate's website offers incredibly detailed and easy to understand explanations, graphs and resources to help the general public grasp the causes, impact and projections. The report paints a dire picture for the land *down under*.

To my fellow Australian readers, you'd probably be across most of these projections given recent media attention. These include more extremely hot days and fewer extremely cold days, ongoing sea level rise, further warming and acidification of the oceans around Australia, more frequent and severe bleaching events on the Great Barrier Reef, more time spent in drought in many regions of southern Australia, short-duration extreme rainfall events throughout Australia, longer fire season for southern and eastern Australia, and a greater proportion of high-intensity storms. People are already experiencing these effects firsthand, both in Australia and around the world. In the US, for example, the number and intensity of hurricanes has been increasing since the 1980s.

While writing this chapter, the latest news from the UK came past my desk (in my Google news feed) regarding the respected English broadcaster and natural historian Sir David Attenborough having been invited to speak to the UK Parliament's Business, Energy and Industrial Strategy Committee on a range of issues relating to climate change and the net zero emissions target. The video showed a dismayed Sir Attenborough answering a question about people in positions of power who were sceptics. He said he was sorry there were such people in power around the world, particularly in Australia, "because Australia is already facing having to deal with some of the most extreme manifestations of climate change".

STEMM LEADERSHIP OR LEADERSHIP IN STEMM?

Many individuals will initially study a field in STEMM and somewhere along their careers choose to pursue a different path, outside their original area of interest. For instance, as we saw in chapter 1, of the top 100 CEOs celebrated by *HBR*, 34% have engineering qualifications. Other studies seem to suggest that CEOs with STEMM undergraduate studies outperform their non-STEMM peers in STEMM industries such as pharmaceuticals and technology. Such leaders transitioned from their respective STEMM fields and became business leaders. In other examples they've gone on to become international leaders (see figure 12). This is what we call "STEMM Leadership", and each individual is referred to as a "STEMM Leader". These are individuals who have leveraged the foundation provided by a STEMM education, and perhaps even a STEMM career, to go on to drive systemic change.

Figure 12: STEMM Leaders: Margaret Thatcher, Angela Merkel and Bill Gates

Images: Shutterstock.com

For instance, Margaret Thatcher studied law at Oxford after completing a four-year Chemistry degree and working in the scientific field, albeit for a short time. She was reportedly prouder of becoming the first Prime Minister with a science degree than becoming the first woman. Angela Merkel has a doctorate in Physical Chemistry and was actively involved in various political activities before being elected to office. Bill Gates is a software developer who, despite dropping out of Harvard, went on to become one of the most accomplished businesspeople of our time after co-founding Microsoft and later becoming a powerful advocate for global health and development.

Systemic change is required

Systemic change is required when efforts to change one aspect of a system fail to fix the problem. The whole system needs to be transformed. Systemic change can mean gradual institutional reforms, but those reforms must be based on and aimed at a transformation of the fundamental qualities and tenets of the system itself. When our objective is systemic change, we need to look at the whole system, including all its components and the relationships between them. Chapters 1 and 2 provided the reader with greater context regarding the system in which STEMM professionals operate – a bird's eye view, if you will. Future chapters will address factors from closer (and multiple) perspectives.

To be sure, we need STEMM experts (that is, leaders in STEMM and leadership in STEMM) to continue their remarkable work in their chosen fields. As discussed in chapters 1 and 2, the world desperately needs the continued innovative solutions and discoveries made by STEMM professionals. This is pivotal to achieving the UN Global Goals, mitigating global risks and transforming the world for the good of everybody in the Fourth Industrial Revolution. For this, we need to encourage more young people to study and pursue STEMM careers. We also need to create the right organisational conditions to motivate them to remain in the STEMM workforce – according to the Pew Research Center, in the US only 52% of STEM-trained university graduates are employed in the STEM workforce (medicine was not included in the survey).

To be clear, this is not about STEMM professionals abandoning their areas of expertise to pursue careers in politics, become board members or CEOs, or start new companies. Conversely, this is not about replacing non-STEMM leaders with STEMM leaders. One is not better than the other. Instead, this is about achieving a new leadership paradigm in response to the needs of 21st century organisations and the world at large in the face of the Fourth Industrial Revolution. Just like any resilient system (team, organisation, city, state, nation) must adapt to new challenges with agility, so do our leadership paradigms.

In a democracy, we need **cognitive diversity** within leadership circles (that is, different thinking styles, knowledge, skills, values and beliefs). We need leaders who can bring multiple perspectives to the decision-making table and avoid the pervasive "groupthink" psychological bias. That is, current leaders often seem to prefer harmony or conformity in the group, which results in an irrational or dysfunctional decision-making outcome. As history has shown, groupthink can produce dehumanising actions against the "out-group". Furthermore, extensive research has shown that cognitive diversity in teams leads to greater creativity, faster problem-solving and better outcomes.

THE STEMM LEADERSHIP ADVANTAGE

This is a deliberate attention-grabbing heading to pique the reader's curiosity. After all, "STEMM leadership advantage" would imply that STEMM leaders are superior to other leaders, which as was just explained, is not the message of this book. To suggest superiority would represent the opposite of our earlier plea for diversity and collaboration, where everyone can be represented and contribute. There are, however, very distinct characteristics of STEMM minds that beg to be highlighted. What makes STEMM minds different may in fact be an advantage to organisations and the rest of the world.

Teri Odom, Professor and Chair for the Chemistry Department at Northwestern University (Illinois, US) and an alumnus from Stanford and Harvard, wrote an article in 2015 for *The Huffington Post* with another attention-grabbing title: "Scientists Make Better Leaders". Like many other heads of science departments in Australian universities, she expressed concern over the dwindling numbers of students choosing to study STEM fields. Professor Odom believes that training in science represents a great return on investment for anyone considering any kind of career where **analytical and complex problem-solving skills** are required. This is because the scientific approach to problem solving has its roots in **critical and analytical thinking, teamwork and creativity**. She went on to say:

In science, when we first learn something, we may not yet understand that there are often many different solutions that can produce the right answer. The goal in science is not to be right, but to **make discoveries that push our understanding of how things work**, which can be tested in multiple ways and be reproduced by others. The goal is to **reveal knowledge for the common good**. And at its core, the scientific method is driven by bold ideas premised on hypotheses that may, in fact, end up being wrong. But then, adjustments are based on new data, and a more plausible hypothesis can be formed, then tested and then interpreted as best that we possibly can. **Perseverance and long-term thinking** are critical.

Imagine if those in local and global leadership were trained in this way. Those leaders trained in STEM disciplines would surely have a leg up on those who had little to no exposure to scientific thinking. The world is a complex place. We need more science-trained leaders who will advocate for the common good **in tense and fluid situations**. We need leaders whose **influence extends beyond a narrow group of followers** ...

(Bold added by author for emphasis.)

The "STEMM leadership advantage" consists of those characteristics and qualities cultivated from an early stage within STEMM education and professions, and which are sorely needed in today's world. Professor Alice Gast – President of the Imperial College London, past US Science Envoy, and a chemical engineer from Princeton – added a few more characteristics in her WEF article "Why Business Leaders Should Think Like Scientists". These included **sceptical curiosity** (scientists need to be sceptical to minimise the effects of biases and arrive at the most accurate solutions); **collaborative competitiveness** (collaboration and data sharing are the purview of "big science" projects); and **confidence in the face of uncertainty and the unknown** (the scientist's business is the unknown).

By all accounts, having more leaders from STEMM fields in influential roles would appear to be a very compelling proposition. Indeed, a not-for-profit organisation called 314 Action was founded in 2016 by chemist researcher Shaughnessy Naughton, who ran for Congress in Pennsylvania. The organisation's mission is to connect people with STEMM backgrounds with the skills and funding needed for a successful political campaign for public office in the US. Clearly, people are starting to recognise the urgent need for more diverse leadership, which has been reviewed in this book thus far. We've also reviewed the existing gaps in today's leadership circles; we've identified several highly successful STEMM professionals who went on to become influential world leaders; and we've articulated their characteristics and qualities, which are widely regarded to be essential for the future and wellbeing of our world. This then leads to the question:

Why is it that we don't see more STEMM leaders making it to the top?

To answer this question, it's best to consider it from a coaching psychology perspective; instead of asking *why* (which may take us down a rabbit hole and hinder the identification of potential solutions), reframe the question using *what* and *how*. As a leadership coach working with STEMM and non-STEMM leaders within the organisational context – that is, from a systems perspective – I always start from the premise of "what needs to happen for the individual to flourish within the organisational context?". As such, let's ask the following questions, with the view to seeing more STEMM professionals in the boardroom, C-suite and other highly influential positions:

- What skills and capabilities does the STEMM employee(s) need to develop to become an effective leader(s) within their organisation(s), and how can they go about it?

- What can CEOs, HR leaders and coaches do to help their STEMM employee/client reach their full potential, including leadership opportunities?

- How can STEMM employees learn and adapt to a wider range of communication styles within a commercial context?

- How can STEMM and non-STEMM employees develop greater self-awareness, emotional intelligence, resilience and psychological flexibly?

- What systemic changes need to be made within teams, organisations (and society) to achieve greater cognitive diversity? How can they be achieved?

This brings us to The SCIENCE of Leadership, a coaching model that will hopefully illuminate potential answers ...

PART II

THE SCIENCE OF LEADERSHIP MODEL

CHAPTER 4

A NEW APPROACH TO DEVELOPING LEADERS

"Leaders are made, they are not born. They are made by hard effort, which is the price which all of us must pay to achieve any goal that is worthwhile."

Vince Lombardi

HOW THINGS CAN GO FROM BAD TO WORSE

Let's take a look at a real-life example of the types of challenges we frequently encounter in our coaching practice, to help us understand the complexity of leadership interactions in the workplace involving STEMM professionals.

Gina was an R&D Director with a PhD in Biochemistry in her early thirties who had moved up the corporate ladder faster than was customary in most STEMM fields. One year, her employer – the Australian subsidiary of one of the largest pharmaceutical multinationals globally – decided to invite other departments to the two-day national sales conference with the view to bolstering multidisciplinary collaboration on innovation projects. This seemed like a great idea. However, Gina felt drained and restless at the end of the conference. Back at the office, a number of pre-existing problems began to surface within her team and with other parts of the business; for example, conflict (emails and meetings), an increase in the number of sick days taken, missed project deadlines, and the late launch of an important new product.

To help figure out what was going on, I was contacted by Geoff, the VP of Product Development (based in the US) and a highly successful scientist prior to becoming an executive, whom I'd known for 20 years. Geoff was aware of my focus on coaching STEMM leaders. Gina, the HR Manager, the Australian Managing Director and I had a few conversations (individually and jointly), and after all parties agreed to a plan, Gina and I began working together. I would be helping her bolster her leadership capabilities through our coaching program for senior leaders.

As R&D Director, Gina was respected as a scientist and well regarded as a leader of people. The company had sent her to a few external leadership courses to support her transition from manager to director a couple of years prior. The 360-degree feedback (interviews and reports) showed that Gina was effective at negotiating with government regulators, technical service providers and leading her team. She was especially effective at coaching her scientists to work cohesively and collaboratively among themselves, and with the Quality and Production departments. Her R&D department had scored the highest possible rating at an international quality audit led by headquarters based in the US. However, the company was meeting only 90% of new product development targets, for which R&D seemed to be held largely responsible.

As Gina became more comfortable and began to feel "safe" in our coaching sessions, we started to get to the heart of the problem. She candidly shared her thoughts and feelings about an event she believed had adversely affected her and her team, resulting in dire results for the company. There had been a facilitated workshop on the first day of the conference, led by one of the company's HR managers, where everyone was asked to share their Myers–Briggs Type Indicator (MBTI) results. Throughout the next two days, people were calling each other by their MBTI type, accompanied by body language signalling admiration and approval, and at the other end of the spectrum, pity or disregard. Gina took note of the patterns, and how a large portion of her team appeared to be pitied or disregarded by their sales and marketing colleagues. Many even approached her with utter disbelief that she could be an introvert. The labels stuck ("Hi INTJ, how are you today?"), even after several months.

Testing based on subjective observation

MBTI is perhaps the most popular personality test, used by about 80% of Fortune 500 companies, and it's especially pervasive in the pharmaceutical industry. The test uses a series of yes–no questions to break down personalities along four dichotomies:

- Introversion (I) vs Extroversion (E)
- Sensing (S) vs Intuitive (N)
- Thinking (T) vs Feeling (F)
- Perceiving (P) vs Judging (J).

The test generates a four-letter combination which supposedly represents a person's psychological "type". The HR manager had grouped the results according to type on a whiteboard and asked people to comment on the positives and negatives of each type, based on descriptions and "best careers per type" information previously circulated.

Many social biases surfaced during that process, as those with an "I" for introvert would get snickering and condescending remarks from the extroverts in the room, as if being an introvert was an illness to be cured or a defect to be fixed. Similar remarks were made towards those with an "F" for feeling, because the perception within large corporates is that to be successful one must be tough and not emotional. Granted, the HR manager tried (which Gina thought was not genuine) to clarify that the purpose of the exercise was for everyone to become aware of individual differences and learn to work more collaboratively by modifying their own behaviour. However, her exhortation was very brief and lacked "how to" substance. Over time, the R&D team started to use passive aggressive behaviours in retaliation; for example, unanswered emails, no-shows and late meeting arrivals, and short responses.

After a few more coaching sessions, Gina developed the necessary capabilities to have conversations with the Sales and the Marketing Directors about establishing respectful dynamics between all functional teams. Over time, the alienating and judgemental dynamics toward the R&D team subsided. This was aided by workshops on dialogue skills and how to break through gridlock with the power of dynamically authentic conversations. The teams collectively agreed on what unhelpful behaviours looked like (for example, coercion, manipulation, condescension, and passive aggressiveness). Gina's coaching sessions continued separately, as we'll see further on.

Figure 13: Dilbert and MBTI

In the meantime, interactions inside and outside new product development meetings (or NPD meetings) became more effective. Within six months, new product targets improved from 90% to 98% success rate (launched on time and within budget). Another six months later, the company achieved their desired 100% target rate.

The limitations of Myers–Briggs

The limitations of the MBTI are well known among psychology academics. The approach was created by Katharine Cook Briggs and her daughter, Isabel Briggs Myers, in the US in the early to mid-20th century. Briggs was inspired to research personality type theory when she met Isabel's future husband, Clarence Myers. She noticed he had a different way of seeing the world. This intrigued her enough to start a literature review to understand different temperaments. Neither had any formal psychological training, therefore they based their work on their observations and own interpretations. Although proponents of MBTI will point out its link to Carl Jung's *Psychological Types* from the 1920s, the original Jungian concepts are distorted, even contradicted, according to Adrian Furnham, Professor of Psychology at University College London.

In 1991, the National Academy of Sciences analysed more than 20 studies of the MBTI and concluded that the scale had not demonstrated adequate validity and that "there is not sufficient, well-designed research to justify the use of the MBTI in career counselling programs". In Australia, the following statement was published in 1995 in *Australian Psychologist*: "there are several psychometric limitations pertaining to the reliability and validity of the MBTI, which raise concerns about its use by practitioners. In view of these serious limitations, routine use of the MBTI is not recommended, and psychologists should be cautious as to its likely misuse in various organisational and occupational settings."

In 2018, Oxford Associate Professor Merve Emre published the outcomes of her exhaustive research on the $2 billion personality testing industry – specifically MBTI – in her book *The Personality Brokers: The strange history of Myers–Briggs and the birth of personality testing*. She explains how the test is now called an "indicator" as a way of getting around questions of validity. As a humanist, Emre explains, what matters most is the language of "type" – such as the categories of extroversion and introversion – which put people into a box and imply *good* or *bad*, *better* or *worse*. There's also the issue of employers invading employees' privacy, as was the case with the abovementioned example and the common practice of asking employees to openly share their "type".

Scientific and evidence-based reasoning

So, what should companies be using instead? We recommend the use of evidence-based psychometric tests in leadership and team coaching programs for developmental purposes. We always go the extra mile with clients to explore what is meant by scientific and evidence-based reasoning, and encourage CEOs and HR leaders to implement evidence-based management and leadership practices. But this is much easier said than done, as many experienced practitioners would confirm.

THE SYSTEMIC ISSUES HOLDING STEMM LEADERS BACK

During my time as a student in the MBA Executive program, I noticed that almost all references used in the course material came from business journals and magazines. The more established subjects such as economics, accounting, marketing and finance included textbooks, however even those referred to models and concepts that were hard to trace back to primary research data.

I mentioned to my classmates how concerned I was, as a scientist, to be studying organisational strategy and leadership from case studies, management consulting firms, interviews and anecdotes, and that the definition of "research" seemed to have a different meaning in the social sciences, including management and leadership. I could hear the crickets, then I saw the smirks as they all looked at each other – a group of four bankers and a management consultant. The penny dropped: the discovery of the systemic issues holding STEMM leaders back had commenced. We are still dealing with this problem worldwide, as citizens are expected to use information to make decisions about health, behaviour and public policy. But where is this information coming from?

Marketers, advertisers and the media often overstate or simply don't understand the implications of scientific evidence, overlooking methodological or statistical flaws, or even relying on pseudoscience. Poor evidence skills are especially problematic for issues related to education (for example, leadership development), public policy (for example, climate change) and personal choice (for example, anti-vaxxers). The current state of science (explained further on) combines with the nature of scientific writing in the media to add to the confusion we currently observe almost everywhere. This leads to the question: shouldn't it be the goal of the education system to teach students the inquiry skills necessary to critically evaluate scientific evidence and to assess whether evidence is consistent with claims or theories?

Interestingly, a 2018 Pew Research study reported that 52% of adults in the US think that the reason young people don't pursue STEM degrees is they think these subjects are too hard.[8] The problem is not necessarily to do with earning capacity either, since STEM university majors tend to earn more than non-STEM university majors, according to the Pew Research Center. Unfortunately, the problems associated with the challenges surrounding everyday scientific reasoning extend beyond studying STEM at university.

An inescapable reality is that people are highly influenced by their own beliefs when evaluating evidence. In addition, people are "motivated reasoners" and are influenced by their hopes and emotions in addition to their beliefs, especially when the context of the evidence is relevant to decisions in their own lives. When evaluating evidence that's congruent with their own beliefs, there is a tendency to rely on heuristics (mental short cuts) and not engage in analytical thinking. This is known as "confirmation bias". We have a strong tendency to latch on to anything that supports our existing position and blindly ignore anything that doesn't. Beliefs and tendency towards heuristic thinking often go hand in hand. In contrast, belief-incongruent evidence often triggers analytical thinking.

When viewing decision-making through the lens of potential for confirmation bias (our beliefs + our heuristics), it starts to become clear why we must slow down our thinking in order to identify, unpack and evaluate the impact of our belief system, as well as our

---

8   Kennedy, B., Hefferon, M., & Funk, C. (2018). "Half of Americans Think Young People Don't Pursue STEM Because it is Too Hard." Pew Research Center. Retrieved from https://www.pewresearch.org/fact-tank/2018/01/17/half-of-americans-think-young-people-dont-pursue-stem-because-it-is-too-hard/

mental programming, when we're assessing information; that is, evidence.

As a concept, the field of medicine is the gold standard of evidence-based practice. The term "evidence-based medicine" was coined in the 1990s and is defined as "the conscientious, explicit and judicious use of a current best evidence in making decisions about the care of individual patients".

EVIDENCE-BASED MANAGEMENT

Applying the concept of evidence-based decision-making can be challenging in management/leadership practice. Most management and leadership decisions are not based on the best available evidence, as history can confirm. Instead, managers and leaders often prefer to make decisions based on their personal experience. However, personal judgement alone is not a very reliable source of evidence because it's highly susceptible to systematic errors – cognitive and information processing limits make us prone to biases, negatively affecting the quality of our decisions. The challenges are further compounded when we consider large systems; for instance, large organisations, local governments, national governments and international initiatives.

In 2002, Professor Gerard Hodgkinson, Professor of Strategic Management and Behavioural Science at Alliance Manchester Business School, was among the first to write about the trend of moving away from "the scientific inquiry approach" and towards "the problem-solving approach". The problem-solving approach constitutes a more socially distributed form of knowledge production, in which knowledge is generated in the context of application by multistakeholder teams, drawn from a range of backgrounds that transcend traditional discipline boundaries. It results in immediate or short-to-market dissemination.

I had the pleasure of sitting in several lectures by Hodgkinson in 2011 during his professorial visit to UNSW, and subsequently spoke to him at length about various topics related to change management and the quality of decision-making behind strategic and large-scale organisational changes. As I embarked on a new career as a consultant and executive coach, we maintained email contact during which he generously provided several papers on the political challenges surrounding evidence-based management (EBMgt). His work was instrumental in the development of a nuanced understanding of human and systemic impediments to scientific inquiry and rational decision-making.

Figure 14: Problem-solving approach to evidence-based management

Adapted from: Barends E., *In Search of Evidence*, 2015

As figure 14 shows, EBMgt is influenced by several sources, reflecting the complexity and richness of the challenges confronting modern organisations, especially those with highly political cultures. As observed in *The Oxford Handbook of Evidence-Based Management*, in Hodgkinson's chapter titled "The Politics of Evidence-Based Decision Making", all forms of organisational decision-making are inherently political. As such, this ought to be overtly recognised. This recognition may increase the responsibility on managers and leaders to be more transparent about the beliefs and values affecting their decision-making. In being more transparent, the benefits to all stakeholders might be enhanced.

The role of business schools and leaders in evidence-based management practice

This is an opportune time to come back to Gina's coaching (from the beginning of this chapter). During our conversations, we discussed the validity of the MBTI results, and whether they were truly relevant to her leadership development. Furthermore, we explored the systemic factors surrounding the company's choices for using this kind of "intervention" (a term commonly used in psychology to denote the combination of a tool and its application; in this case, the use of MBTI, and the sharing of results in a workshop). The intention behind this exploration was not one of criticism or judgement against the company or anyone, given that when it comes to systemic root causes they're usually not ill intended. To understand what is meant by this, let's unpack it further here.

One of the primary issues of EBMgt implementation is helping people in the organisation understand the best research and then integrate it into their day-to-day practices. As mentioned earlier, leaders, managers, HR folk and consultants (that is, practitioners) make decisions largely based on a combination of their own experience, cultural precedents, organisational procedures and policies, and internal data. In addition, it's generally known that any work involving interpersonal interactions, as is the case with managerial/ leadership work, differs from clinical work (the gold standard of evidence-based practice) significantly. Unlike medicine or nursing, management/leadership is not a profession, which means there is no agreed and shared knowledge base.

In a seminal paper in 2005, Professor Denise M. Rousseau – a Professor at Carnegie Mellon University, Fellow of the American Psychological Association, and the originator of the "psychological contract" concept between employer and employees – stated that the most important reason EBMgt is not a reality is not due to leaders/managers themselves or their organisations. After all, in today's hyper-competitive and fast organisational environments, it would be near impossible to expect leaders, including HR, to work against time pressures, resource constraints and political obstacles in the name of scientific evidence. Instead, Rousseau went on to admit that academics like her, and those in similar programs, must accept a large measure of blame.

That's because research evidence is not the central focus of study for undergraduate business students, MBAs or executives in continuing education. And just like I realised at the watershed moment with my MBA colleagues during a team meeting, Rousseau points out that case examples and popular concepts from non-research-oriented magazines take centre stage instead. There are no communities of experts vetting research on effective management/ leadership practices. Few MBAs encounter a peer-reviewed journal during their student days, let alone later. Upon graduation, few business alumni recognise that the knowledge they have acquired will be surpassed over time by new findings. Although social science knowledge continues to expand, business school training does not prepare graduates to tap into it. We end up with managers/ leaders without clear ideas of how to update their knowledge as new evidence emerges.

These are some of the reasons why MBTI continues to be the prevalent psychometric tool in the corporate world, despite having been discredited by the likes of the Australian Psychological Association, the National Academy of Science, and by Wharton School at the University of Pennsylvania. Luckily, with a stellar background in scientific research and with 10 years of experience in organisational leadership, Gina was able to understand the implications of such a systemic issue, as we explored it together. It helped validate her concerns and helped her be open to the leadership program she had embarked on which, funnily enough, was anchored in evidence-based practice.

COACHING AND EVIDENCE-BASED LEADERSHIP DEVELOPMENT

Bill Gates began his 2013 TED Talk with the statement: "Everyone needs a coach. It doesn't matter whether you're a basketball player, a tennis player, a gymnast or a bridge player." Coaching is now viewed as key in leadership and development programs within large organisations, according to C-level executives at the International Coaching in Leadership Forum (ICLF) held at the Harvard McClean Medical School Institute of Coaching (IOC) in 2018. Represented at the ICLF were C-level leaders from Intel, the World Bank, PayPal, WD-40, the US Federal Government (from previous and current US administrations), Tokidoki, Corent Technology and Harvard Kennedy School, among others. They were unanimous that coaching was an essential aspect of their leadership success.

Studies show that executive and organisational coaching is becoming a pivotal component of evidence-based learning and development programs in Australia and globally because business leaders at all levels (board of directors, C-suite executives, human resources, directors and managers) need more sophisticated and adaptive human capital solutions to help them deliver on their vision and achieve organisational objectives sustainably. Executive and organisational coaching is such a solution to the flexibility, confidentiality, safety and real-time solutions focus offered to its participants.

In addition, the return on investment (ROI) has been ascertained in many studies, not only in financial terms to both the individual (for example, promotions) and the business (top- and bottom-line gains) but also in terms of improved employee engagement and wellbeing (see figure 15).

Coaching is a forward-moving, solutions-focused process or endeavour that is collaboratively created by the coach and the client (an individual or a team). During this process, the necessary conditions for learning, growth, and change are explored, and ultimately outcomes are achieved based on the needs of the client.

As a goals-focused endeavour, coaching provides the ideal support to leaders and organisations. Furthermore, coaching has become a preferred framework for progressive organisations that have chosen to phase out traditional performance management systems and have instead implemented a more collaborative approach.

Figure 15: The benefits of coaching

22%	IMPROVED PROFITABILITY
32%	IMPROVED STAFF RETENTION
39%	IMPROVED CUSTOMER SERVICE
52%	IMPROVED CONFLICT MANAGEMENT
61%	IMPROVED JOB SATISFACTION
67%	IMPROVED TEAMWORK
77%	IMPROVED LEADER-FOLLOWER RELATIONSHIP

Source: ProVeritas Group brochure, 2019

In Australia, coaching within the workplace context is generally known as Executive & Organisational Coaching, per the Standards Australia handbook *Coaching in Organizations*. The handbook's development was spearheaded by Professor Michael Cavanagh, from the Coaching Psychology Unit at the University of Sydney (the home of the world's first tertiary course in coaching). The term "business coaching" is also common. We view individuals, teams and organisations from a systems-thinking, and therefore, a holistic perspective. An individual is unique by virtue of the interrelatedness of the parts of his/her system, which come together in a time and context (for example, marketplace or organisational reality, geography, or family setting).

In a co-created coaching space by the client and the coach, **the coach helps the client understand how factors interact and impact behaviour**. All conversations are underpinned by the humanistic psychological principle of unconditional positive regard for the client, where a behaviour is isolated from the person who displays it. Confidentiality and psychological safety must always be maintained. **The coaching process is not linear; it is iterative and highly fluid**, during which the coach draws from several critically researched theories and tools, depending on the individual's needs. The coach drives the coaching process over several sessions (one on one for individuals, and group sessions for teams). The main evidence-based models and tools used in coaching psychology are:

- Systems Theory
- Positive Psychology
- Cognitive Behavioural Theory (CBT)
- Psychodynamics
- Adult Development and Learning
- Self-determination Theory
- Personality Science
- Mindfulness
- Acceptance and Commitment Therapy (ACT)
- Neuroscience and Leadership Studies.

What to look for in a leadership coach

Does your coach know how to create tipping points?

Manfred Kets de Vries, a world-renowned leadership and organisational coach and author, entered a helping profession because of a life-changing event. He was born in occupied Holland during the Second World War, a period of immeasurable human tragedy. This unimaginable suffering ignited his interest in why some people in a leadership position abuse the power that comes with the job. Kets de Vries has made it his life's work to uncover the terrible consequences of dysfunctional leadership, while bringing back the human dimension to organisations.

While living through such dark times in history would be an unrealistic prerequisite, stories of overcoming adversity seem to propel some individuals to help others. Perhaps, when hiring a leadership and team development coach, it might be worthwhile keeping in mind a different mental framework: the "twice born". This term was coined by the Harvard psychologist and philosopher William James, and it is used by Kets de Vries in his 2014 book *Mindful Leadership Coaching: Journeys into the interior* when explaining how overcoming adversity may transform a person to become a change agent.

James and Kets de Vries suggest that when individuals follow a linear life journey and become successful, they are tied to familiar territories and tied to feeling comfortable. But when people undergo a major upheaval and still manage to flourish, it's because they transcended their self-limitations and discovered creative ways of dealing with adversity. Twice-born individuals are hence said to have a different way of relating to other people and the world around them because of their deep recognition of the fragility of life.

It could very well be that, underneath all the empathy, self-ease and confidence projected by great coaches, there are compelling life stories about overcoming adversity, about forgiving, achieving deep learning and self-awareness, and ultimately, reinvention. This makes the coach relatable and … human. As studies suggest, one of the most important determinants of successful coaching is the relationship between the coach and the client; that is, the coaching relationship or alliance. Do you want a coach who is going to be only a cheerleader (yes, we need that, but not as the *only* characteristic)? Or do you want someone who has the knowledge, experience, empathy, strength and courage to know when to push and pull back and create "a-ha" moments? These are the moments that lead to transformation and growth and are also called "tipping points" by Kets de Vries.

These "a-ha" moments don't come out of the blue; they grow out of hours of thought, reflection and preparation on the part of the coach. Although the visible effects of deep change may seem out of the ordinary, they're the result of many small preparatory interventions. Small changes combine to produce big changes since they lay down new cognitive, emotional and behavioural pathways. The visible effects however seem fast and dramatic because the client hasn't seen all the work going on behind the scenes before deep change occurs.

Do they have the right training and experience?

We didn't always have guidelines for executive coaching, let alone consistency of practice, as described in the seminal *HBR* article "What Can Coaches Do for You?", published in 2009.[9] A salient point made by one of the article's contributors, and the pioneer of coaching psychology, Professor Anthony M. Grant, was that executive coaches ought to have formal training in mental health issues. Coaches would be able to detect such issues and help the client achieve mental strength. Grant based his suggestion on the evidence of a 2009 study by The University of Sydney which found that 25% to 50% of those seeking coaching had markedly high levels of anxiety, stress and depression.

The executive coaching field has evolved significantly since the inception of the world's first Coaching Psychology Unit at the University of Sydney in 2000. There are now many academic institutions worldwide offering courses in coaching science, which not only benefits the coaching discipline as a bona fide profession steeped in psychology, but more importantly, it benefits the clients that coaches are meant to help. Furthermore, coaching science has become a galvanising force between disciplines that historically had operated in silos; for example, psychology, leadership/management theory and practice, neuroscience, medicine and socioeconomics. The efforts of the Institute of Coaching (IOC), Harvard Medical

---

9   Coutu, D., & Kauffman, C. (2009). "What Can Coaches Do For You?". *Harvard Business Review*. Retrieved from https://hbr.org/2009/01/what-can-coaches-do-for-you.

School, in driving a scientific approach to coaching practice in the US and internationally are worthy of note.

It's important for potential clients and organisations to look beyond credentials, certificates and degrees and ensure that the coach(es) they are vetting possess formal training in coaching sciences (steeped in psychology) and any other formal qualification and/or experience relevant to the client's reality. After all, individuals are putting their careers, their relationships at work and at home, and their ability to develop the necessary capabilities to flourish in the 21st century in the hands of the coach.

Do they belong to a global community of coaches?

Coaches who study the coaching sciences are taught a variety of evidence-based tools to design coaching interventions that motivate individuals to learn, change, grow, and succeed. In addition, when using coaching interventions underpinned by scientific research, clients will achieve the best possible outcomes. Just like any other professional field anchored in science, coaching will continue to evolve as more research studies are conducted globally. Collaboration between academia and practitioners continues to increase, along with the growing prominence of major professional associations.

While coaching continues to be a self-regulated field, there are global bodies like the Association for Coaches and the Institute of Coaching, Harvard Medical School, which provide evidence-based training, professional codes of practice and ongoing development. Both organisations focus on scientific research, evidence and methodology for coaches working in organisations, either internally (as an employee of the organisation) or externally (as a hired coach).

When considering hiring an executive coach, whether you are an HR leader, a CEO or an individual seeking to advance at work, it is important to look beyond the marketing material and ensure the programs are evidence-based. The right executive coach can often make all the difference in someone's career, and usually in all other aspects of life. Being discerning up front will result in the selection of an executive coach with the highest degree of professionalism and someone who is committed to the client's coaching journey, no matter how challenging it may be.

STEMM AND NON-STEMM LEADERS: WHAT'S THE DIFFERENCE?

Before considering the SCIENCE of Leadership model, it's important to address one of the most interesting questions many executive coaches, including me, have encountered: "Is there a difference between coaching STEMM and non-STEMM leaders?" As the reader might expect, the answer is yes ... and no. This is where humanists, personality scientists, business and HR leaders, neuroscientists, human rights and diversity advocates, and even coaches are likely to have differing views. The answer to that question depends on the perspective taken.

In her 2009 paper "Development Coaching: Helping Scientific and Technical Professionals Make the Leap into Leadership", Dr Jean Hurd, a US-based executive coach, starts by asking why it is so difficult for many technical professionals to become effective managers. A VP of Quality is quoted as saying: "one day we should write a book about why it's so hard for technical people to become effective managers". This paints a less-than-positive picture about STEMM professionals, it seems. The article goes on to identify four main characteristics that are key to being successful as a scientist "but can present challenges in their transition to a leadership role":

1   A view of science as a calling, the core of their identity.

2   Independence in thought and action.

3   Love of creativity and exploration.

4   Technical expertise.

And yet, as we just learnt from recent government reports, and large independent bodies like the UN (including its agencies like UNESCO) and the WEF, these are some of the skills the world is in great need of. Hence, the global drive to improve STEMM education and improve career opportunities. To be fair, perhaps Hurd's work may be more relevant to the US multinational pharma context. However, as you'll read down the track, it fails to address the larger systemic issues that my client Gina encountered. Nowadays, organisations are emphasising diversity and inclusion.

Much has changed in just a few years, and we now know that if we wish to achieve innovative, productive, safe, transparent and positive organisations, we must embrace cognitive diversity and inclusion in the workplace. We must be mindful of everyone's unique strengths, respect individual differences, and create an environment whereby everyone can contribute towards common goals. That is the higher calling of a leader.

Hurd's work – as well as that of others encountered throughout the three-year research period leading up to this book – provides helpful suggestions on how to coach *anyone* who is currently a technical expert or manager and seeking to move up the corporate ladder. This includes accountants, doctors (granted, doctors are part of the STEMM mix), lawyers, actuaries, architects, journalists, graphic designers and engineers (also part of the STEMM family).

Anyone who has achieved expertise in their chosen field is a technical expert, and consequently they would have acquired skills and competencies to get to the top of their game. One of the hallmarks of leadership is the ability to make sense of, and to help others make sense of, complexity. This means that suddenly the newly minted leader must communicate effectively with stakeholders outside of their area of expertise; that is, they need to potentially learn new terms, new communication rituals, new do's and don'ts. This, in fact, is challenging for *everyone*.

THE SCIENCE OF LEADERSHIP MODEL

The SCIENCE of Leadership model (see figure 16) is an approach to developing and coaching leaders that draws from the latest evidence-based theories and practices in coaching psychology internationally. Each of the remaining chapters in this book is based on one of the elements of this model.

Figure 16: The SCIENCE of Leadership model

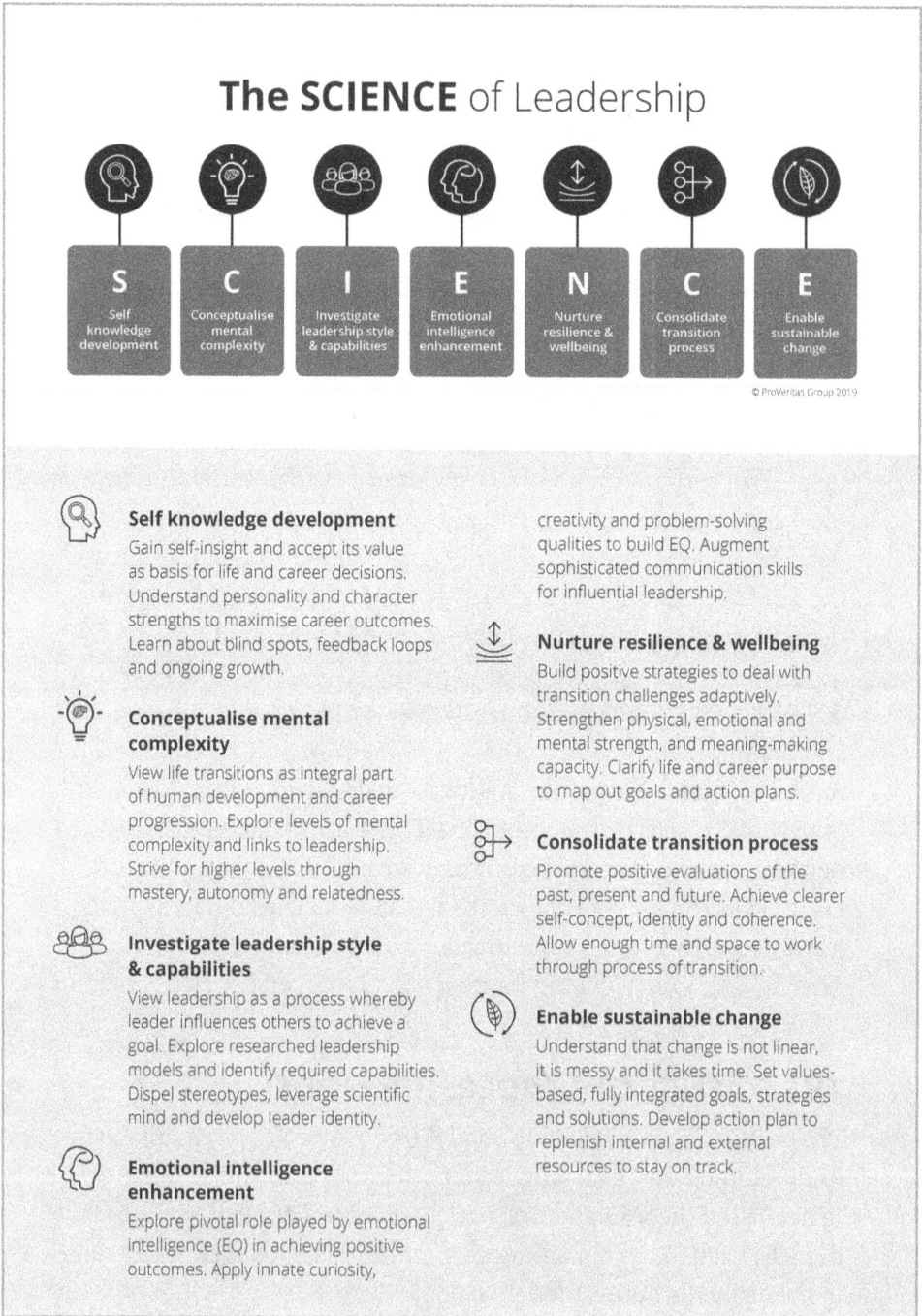

The SCIENCE of Leadership

S	C	I	E	N	C	E
Self knowledge development	Conceptualise mental complexity	Investigate leadership style & capabilities	Emotional intelligence enhancement	Nurture resilience & wellbeing	Consolidate transition process	Enable sustainable change

© ProVeritas Group 2019

Self knowledge development

Gain self-insight and accept its value as basis for life and career decisions. Understand personality and character strengths to maximise career outcomes. Learn about blind spots, feedback loops and ongoing growth.

Conceptualise mental complexity

View life transitions as integral part of human development and career progression. Explore levels of mental complexity and links to leadership. Strive for higher levels through mastery, autonomy and relatedness.

Investigate leadership style & capabilities

View leadership as a process whereby leader influences others to achieve a goal. Explore researched leadership models and identify required capabilities. Dispel stereotypes, leverage scientific mind and develop leader identity.

Emotional intelligence enhancement

Explore pivotal role played by emotional intelligence (EQ) in achieving positive outcomes. Apply innate curiosity, creativity and problem-solving qualities to build EQ. Augment sophisticated communication skills for influential leadership.

Nurture resilience & wellbeing

Build positive strategies to deal with transition challenges adaptively. Strengthen physical, emotional and mental strength, and meaning-making capacity. Clarify life and career purpose to map out goals and action plans.

Consolidate transition process

Promote positive evaluations of the past, present and future. Achieve clearer self-concept, identity and coherence. Allow enough time and space to work through process of transition.

Enable sustainable change

Understand that change is not linear, it is messy and it takes time. Set values-based, fully integrated goals, strategies and solutions. Develop action plan to replenish internal and external resources to stay on track.

The SCIENCE of Leadership endeavours to provide a flexible model for coaches working with any individual undergoing any career transition in management and leadership. It draws from the highly respected INSIGHT model by psychologists Professor Stephen Palmer (President of the International Society of Coaching Psychology) and Sheila Panchal (Director at the International Society of Coaching Psychology). It's also applicable to all senior leaders transitioning to C-level roles, as well as board positions. Since this book focuses on "harnessing the power of STEMM leaders in an irrational world", subsequent chapters will focus on STEMM professionals transitioning to senior leadership roles, incorporating aspects of the SCIENCE of Leadership model.

The SCIENCE of Leadership model is not a sequential process. The steps can be applied in any order deemed appropriate (except, perhaps, for the first one: self-knowledge development).

CHAPTER 5

THE **S**CIENCE OF LEADERSHIP

SELF-KNOWLEDGE DEVELOPMENT

"Make it thy business to know thyself, which is the most difficult lesson in the world."

Miguel Cervantes

In ancient times, people from all over Europe travelled to Greece to visit the Oracle of Delphi, seeking advice on matters of love, war, and commerce. Inscribed above the entrance were two simple words: "Know Thyself". This may sound very simple at first, especially if the person is only a child or if their job doesn't entail influencing others to achieve a goal. Young people, especially, will be adamant that they know who they are, as parents of teenagers will attest. Their perspectives might be more egocentric as they compare themselves with other teenagers in their peer group, a process known as "social comparison". Adults, on the other hand, will have a different definition of self-knowledge as an understanding of their personality, strengths, weaknesses, goals, values, motives, beliefs, and more.

When leading others, we may be confronted with the possibility of not knowing ourselves all that well. Therefore, when working with clients transitioning to senior roles, we spend time in the "self-discovery phase", which initially consists of evidence-based psychometric tests to glean an understanding of personality strengths, values, motives, cognitive ability, and potential behavioural derailers when under extreme stress. We can also test for resilience, mental strength, and psychological flexibly. The information becomes part of the client's self-knowledge once they've given it meaning, made sense of it. This can take many coaching conversations, periods of self-reflection, and more feedback gathering. Indeed, it's an ongoing process because we are always changing in some way or another.

If we recall Gina as an example, the Director of R&D in chapter 4, prior to leadership coaching, she had been pigeon-holed as an INTJ. The feedback within her team was nothing short of stellar, however, the company's 360-degree feedback had uncovered behaviours she seemed unaware of, particularly when dealing with non-technical colleagues. This lack of self-awareness had proven to be detrimental, until we started to explore it within the safety of the coaching space and with the aid of evidence-based tools. Let's read on …

CULTIVATING SELF-AWARENESS

Developing self-awareness is a long and exciting journey of self-discovery, during which leaders uncover who they are and what they want to accomplish, and seek out and value others' opinions. Daniel Goleman (psychologist, journalist and author of *Emotional Intelligence*) suggests that leaders with strong self-awareness have an accurate sense of their strengths and limitations, which gives them realistic confidence. It also gives the leader clarity on their values and sense of purpose. As such, they can be candid and authentic, speaking with conviction about their vision.

According to organisational psychologist Dr Tasha Eurich, in her 2017 book *Insight: The power of self-awareness in a self-deluded world*, most people believe they are self-aware, however only 10% to 15% truly are. This is not at all surprising to those of us working in leadership coaching, especially because most leaders are promoted due to their excellent technical skills, and not due to their ability to understand their own emotions and cognitions (thoughts) or how they help or hinder what they're trying to accomplish. Eurich's research showed that people do not always learn from experience, and that expertise does not help people root out false information. Indeed, highly experienced leaders often won't do their homework in leadership development settings and tend to be more prone to confirmation bias (hence, they won't question assumptions).

Learning to ask the right questions

Similarly, the more power a leader holds, the more likely they are to overestimate their skills and abilities, as shown by several studies. In fact, across a variety of roles and industries, it was found that higher level leaders overvalued their skills (compared to others' perceptions) in competencies like emotional self-awareness, empathy, trustworthiness, and leadership performance. We found this to be the case with both Rob (recall the IT Sales Director from chapter 1) and with Gina, despite their commitment to introspection in the past. They were therefore shocked to learn that studies suggest that people who are introspective can often be less self-aware and report less job satisfaction and wellbeing. The reason for this? Most people do introspection *incorrectly*. Instead of asking themselves "why" they did *this* or *that* wrong, leaders should ask "*what*" questions.

In coaching, we know it's counterproductive to ask "why" questions. They send the client on a wild goose chase looking for reasons, and they generally settle for the easiest and most plausible answer – a result of confirmation bias. Asking *why* can often cause clients to fixate on their problems and place blame, instead of moving forward in a healthy and productive way. Asking why reduces satisfaction with the choices made and can affect mental health because it puts the client into a victim mentality. Instead, it's better to ask "*what*" and "*how*" questions. For instance, when I asked Gina, "*what* are the steps you need to take to build collaborative relationships between the commercial and the technical sides of the business?", she felt empowered to come up with several actions. We then assessed those actions (the *how*) in terms of additional resources needed, priority, and impact. It triggered "realistic possibilities", and if something was outside her circle of control or influence, we would shelve it.

A useful table we like to use in our coaching programs is the 2 × 2 Self-Awareness Archetypes Map, developed by Eurich (shown in figure 17), which differentiates between two types of self-awareness: *internal* and *external*.

Internal self-awareness represents how clearly we see our own values, aspirations, thoughts, feelings, behaviours, strengths, and weaknesses, and their impact on others. The higher the internal self-awareness, the higher the job and relationship satisfaction, and personal and social control. And the lower the anxiety, stress and depression a person might experience.

External self-awareness means understanding how other people view us in relation to the same factors just listed. Leaders who know how others see them are more skilled at showing empathy and taking others' perspectives, leading to better relationships.

Rob, the IT Sales Director, seemed to struggle for a while with the concept of external self-awareness as he was adamant that his "success to date" was due to his personal attributes, and that was the reason he had been hired by this American IT multinational. Rob seemed to unconsciously believe in the trait definition of leadership (shown in figure 6). It wasn't until we obtained 360-degree feedback data that he was able to internalise (make sense of) how his behaviour had become too pushy in the new organisational context. He had been stuck in the "pleasers" quadrant of the 2 × 2 Self-Awareness Map. We needed to go back to initial coaching sessions about his values and *what* truly mattered to him and ask him *what* behaviours were in alignment with those values, thereby moving to the "aware" quadrant. Over time, we worked on becoming truly aware (that is, to be both internally and externally self-aware) through practical exercises, ongoing feedback discussions and detailed discussions about emotional intelligence, as we'll see in chapter 8.

Figure 17: The 2 × 2 Self-Awareness Archetypes Map

	External self-awareness	
	Low	**High**
High (Internal self-awareness)	**Introspectors** Understand who they are but don't challenge their own views by seeking feedback from others. This can harm relationships and limit success.	**Aware** Know who they are and what they want to accomplish, and seek out and value others' opinions. This is where leaders begin to fully realise the true benefits of self-awareness.
Low (Internal self-awareness)	**Seekers** Don't yet know who they are, what they stand for, or how their teams see them. May feel stuck or dissatisfied with their performance and relationships.	**Pleasers** Can be so focused on appearing a certain way to others that they overlook what matters to them. Over time, they tend to make choices that aren't in service of their own success and fulfillment.

Adapted from: Eurich, T. (2018). 'What Self-Awareness Really Is (and How to Cultivate It)' [Blog]. Retrieved from https://hbr.org/2018/01/what-self-awareness-really-is-and-how-to-cultivate-it?

HOW INDIVIDUAL VALUES AFFECT BUSINESS PERFORMANCE

In our leadership coaching programs, we ask our clients to think about a leader who has inspired them, and they'll often mention Mahatma Ghandi, Martin Luther King Jr, Nelson Mandela or John F. Kennedy. When asked what characteristic(s) make them special, the majority seem to sum it up with the word "**character**". Interestingly, there is no universally accepted definition for the word "character". Some define it as the qualities valued by the individual, the qualities that make the person who he/she is, which may be viewed as a combination of his/her virtues, strengths, and talents. The *Cambridge Dictionary* defines it as the combination of qualities, or positive traits, in a person that makes them different from others. The *Merriam-Webster Dictionary* offers this definition for character:

the complex (or sum) of mental and ethical traits marking and often individualising a person, group, or nation.

Prior to 2004, there was no common vocabulary for discussing measurable positive traits. Psychologists Martin Seligman and Christopher Peterson – pioneers in positive psychology – therefore set out to identify, organise and measure character by first defining it as traits possessed by an individual. These traits are stable over time but can still be impacted by setting (situation) and thus are subject to change. Peterson and Seligman examined ancient and present cultures for information about how people in the past construed human virtue. Six core virtues emerged from their analysis, which they viewed as "core characteristics valued by moral philosophers and religious thinkers":

- Courage
- Justice
- Humanity
- Temperance
- Transcendence
- Wisdom.

They then moved down the hierarchy to identify character strengths, which are "the psychological processes or mechanisms that define virtues". Twenty-four strengths – or values in action (VIA) – emerged, which can be assessed using self-report questionnaires, behavioural observations, peer report methods, and clinical interviews. The 24 Character Strengths are shown in figure 18. Positive psychologists have come a long way in terms of evidence-based models, and it's no different with measuring the 24 Character Strengths. Individuals can complete the test (now known as the VIA Survey) online, and they will receive a report ranking the order of their strengths from 1 to 24, with the top 4 to 7 considered "signature strengths".

Figure 18: Summary of VIA Classification of Character Strengths and Virtues

VIRTUE OF WISDOM

Creativity
- *Seeing things differently*
- *Original ideas and concepts*

Curiosity
- *Explorer*
- *Willing to have or develop new experiences*

Judgement
- *Thinks through all aspects*
- *Listens to all sides*
- *Thinks critically*

Love of Learning
- *Likes to learn new things*
- *Masters new knowledge and skills*

Perspective
- *Sees the big picture*
- *Offers guidance and wisdom*

VIRTUE OF COURAGE

Bravery
- *Not afraid to face challenges*
- *Speaks their mind*
- *Holds to what is right*

Perseverance
- *Persistent*
- *Will finish what they start*
- *Problem solver*

Honesty
- *Shows integrity*
- *Sincere*
- *Tells the truth*

Zest
- *Full of energy and enthusiasm*
- *Follows through on everything they do*

VIRTUE OF HUMANITY

Love
- *Gives and receives love*
- *Has genuine relationships*
- *Values others*

Kindness
- *Generous*
- *Provides empathy and kindness*
- *Will help others*

Social Intelligence
- *Aware of what others think or mean*
- *Aware of themselves*

VIRTUE OF JUSTICE

Teamwork
- *Works well in a group*
- *Provides to the group*
- *Has a sense of loyalty and social responsibility*

Fairness
- *Seeks justice*
- *Unbiased*

Leadership
- *Influencer*
- *Can organise a group and activities*

VIRTUE OF TEMPERANCE

Forgiveness
- *Accepts repentance*
- *Forgives and forgets*
- *Shows mercy*

Humility
- *Allows their actions to speak for themselves*
- *Shows modesty*

Prudence
- *Not a risk taker*
- *Careful*

Self-regulation
- *Can manage impulses and emotions*
- *In control*

VIRTUE OF TRANSCENDENCE

Appreciation of beauty and excellence
- *Appreciates aspects of beauty*
- *Admires those with great skill and knowledge*

Gratitude
- *Is thankful*
- *Feels blessed*

Hope
- *Optimistic about the future*
- *Works to achieve positive outcomes*

Humour
- *Makes people smile*
- *Lighthearted*

Spirituality
- *Has faith*
- *Religious*
- *Has purpose*

Your signature strengths are those strengths that best describe the positive aspects of who you are. These strengths are strong capacities in you, and they are probably engaging, energising, and comfortable for you to use. Your family and friends would immediately agree these are important strengths that you have. Finding ways to use and express these strengths can provide benefits and help you flourish at work and in other areas of life. It is important to remember that everyone has the capacity to express and develop all the 24 strengths discussed in this report. In a nutshell, all strengths are important – they all matter – but some are more relevant at certain times than others.

It's difficult to overstate the power of a company with leaders who inspire at every level of the organisation. These are the companies that deliver innovative and even heroic feats in business because so many of the people who work there are motivated to make them happen. Inspired employees are more engaged and are more than twice as productive as simply satisfied employees. The question is, which combination of characteristics or signature character strengths matter most?

Can a leader with a certain combination of four to seven VIA signature strengths be inspirational in every organisation, community, or nation? As most experienced leaders already know, effective leadership isn't generic – it depends on multiple factors. There is no consensus in the leadership literature about the specific qualities, or character strengths, needed to be an inspirational leader.

Becoming an inspiring leader

Excellent performance can only be achieved when a company's leadership profile reflects its unique culture, strategy, business model, and context. In a way, this becomes the company's unique behavioural signature. The same is true for leaders: they need to possess the right combination of signature strengths for the organisational context in which they operate. For instance, a leader whose best talent is cost management will not inspire an organisation that creates value by out-marketing the competition.

Many leaders struggle to understand what is preventing them from flourishing in their leadership roles, which is one reason they may seek executive coaching. As we make progress through the coaching process, and the conversations shift towards the individual's values and strengths, and what is truly important and meaningful, they uncover what truly motivates them. At this stage, they often decide to realign their career and life goals with their values and signature strengths. On the other hand, the person may have the realisation that certain characteristics, or character strengths, need to be cultivated to achieve *a life well lived*, or happier living.

Either way, achieving congruence, or alignment, between the leader's signature strengths and his/her career goals (which need to be aligned with the organisation's goals) would be a precursor to becoming an inspiring leader. This realignment process can be quite nuanced due to the unique history, personality, life stage, and myriad other factors specific to that leader.

LEADERSHIP AND COURAGE

Aristotle said: "Courage is the first of human qualities because it's the quality that guarantees the others". Hundreds of years later, this was echoed by Maya Angelou when she said: "Courage is the most important of the virtues because without it, no other virtue can be practised consistently". Courage is the resolve to act virtuously, especially when it's most difficult. It is acting for the general good, when it would be easier not to.

Courage can be developed, exercised and built

A courageous person understands danger, chooses to overcome their fear, proceeds to face the danger, and acts according to their values. It is not fearlessness, recklessness, or rashness. It is a well-considered decision to behave constructively despite fear, discomfort, or temptation, according to University of Michigan's Law Professor William Ian Miller, in his book *The Mystery of Courage*. It is the discipline to act on wisely chosen values rather than on impulse.

93

Courage can therefore be developed, exercised, and built. This is great news for all of us.

Today's leaders need courage more than ever. We live in turbulent, yet exciting times in terms of STEMM advances. We need courageous leaders to have the difficult conversations that result in change. We need courageous leaders who can understand the risks (especially to their popularity) and will nevertheless choose to overcome their fears to act according to their values. Such leaders inspire hope and future-oriented thinking and are prepared to make wise decisions to create communities of shared purpose.

The latest results from the annual human capital survey conducted by Deloitte in 2019 give cause for pause ... and hope.[10] In it, respondents cited **societal impact** as the top factor used to measure success when evaluating annual performance. This would undoubtedly require a great deal of courage from leaders – courage to make difficult and even unpopular decisions that may challenge the status quo with the view to establishing more sustainable corporate governance practices that benefit everyone.

It probably isn't a coincidence that the theme of this year's global conference of the International Leadership Association (ILA), held in Ottawa, Canada in October 2019, is "Leadership: Courage Required". The ILA is the global network for those who study, teach, and practise leadership. The ILA brings together thousands of leadership professionals from multiple sectors, disciplines, professions, cultures, and generations – to advance leadership knowledge and practice for a better world. I'll be presenting on the topic "Coaching STEMM Leaders to Reach the C-suite and Transform Organisations". It's not a coincidence that I've chosen to present this topic this year. Clearly, we need more courageous leaders to achieve positive societal impact, and they may arise from STEMM leaders if the right individuals and organisational ingredients are cultivated.

---

10   Deloitte Touche Tohmatsu Limited. (2019). "Deloitte Global Human Capital Trends: Leading the social enterprise, reinvent with a human focus". UK: Deloitte Insights. Retrieved from https://www2.deloitte.com/content/dam/insights/us/articles/5136_HC-Trends-2019/DI_HC-Trends-2019.pdf

UNDERSTANDING PERSONALITY TRAITS AND THE EFFECTS ON BUSINESSES

Rob described himself as an introvert, even before we commenced our coaching work together. Gina also offered the same assessment of her personality. Curiously, Rob qualified his description by adding that he was an extrovert at work, and an introvert at home, hence he seemed convinced that he was an ambivert. He explained in our first coaching session that one of the reasons he thought he was struggling to get along with his CEO was that the CEO was a strong extrovert.

These clients had previously undergone an MBTI exercise through their respective companies, and – as explained before – these labels can create a sense of belonging (or sometimes exclusion) and perhaps help you understand yourself and others better. The problem is, labels like these can leave a person with a sense that their personality is fixed and unchangeable; that is, that they are "one way or the other". Personality researchers, however, have a different way of thinking about personality. They focus on *traits* rather than *types*, and use the Big Five personality traits model (also known as the Five-Factor Model, and as OCEAN, as shown in figure 19) because it's the most researched across a wide range of cultures and different demographic groups.

The Big Five model is the result of a systematic process of looking at characteristics used to describe people and putting them through sophisticated statistical analysis. Through lexical and factor analysis, researchers took words that describe personality (approximately 18,000 words), narrowed that list down to include stable traits (approximately 5000 words), factored out synonyms (approximately 180 words), and merged terms related by underlying themes to form five basic dimensions of personality. The five traits are independent and unrelated, and we all have different aspects of each. When measuring someone's traits, psychologists use a spectrum from extremely high to extremely low, rather than a dichotomy like extroverted or introverted.

As we always say to our coaching clients when discussing their psychometric test results, there is no right or wrong answer, or result. Doing the test can be beneficial because it provides us with greater self-knowledge. In turn, this informs our decision to change unhelpful attitudes and modify our behaviours to help us achieve our goals. And for those clients with a keen interest in scientific concepts, we highlight the fact that the field of personality science is forever evolving as researchers continue to challenge and push boundaries. Those boundaries have been well and truly pushed, and indeed are being redefined by neuroscientists. Not long ago, it was thought that the brain you were born with was unchanged throughout your lifespan, and that the brain cells we had at birth were the maximum number we would ever have. The brain was thought to be hard-wired to function in predetermined ways. But that's not true.

Figure 19: The Big Five Personality (OCEAN) model

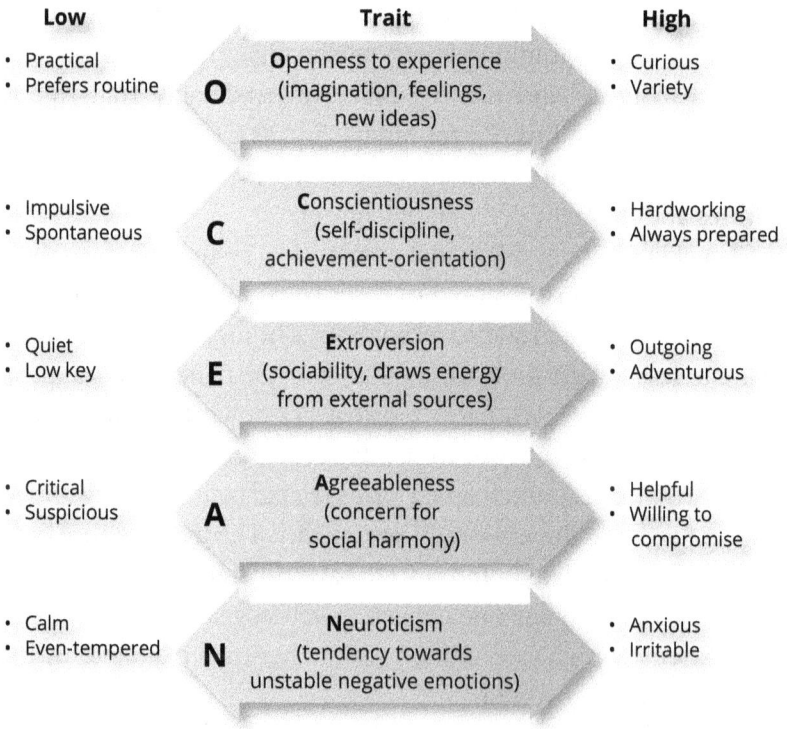

Low	Trait	High
• Practical • Prefers routine	**O** **O**penness to experience (imagination, feelings, new ideas)	• Curious • Variety
• Impulsive • Spontaneous	**C** **C**onscientiousness (self-discipline, achievement-orientation)	• Hardworking • Always prepared
• Quiet • Low key	**E** **E**xtroversion (sociability, draws energy from external sources)	• Outgoing • Adventurous
• Critical • Suspicious	**A** **A**greeableness (concern for social harmony)	• Helpful • Willing to compromise
• Calm • Even-tempered	**N** **N**euroticism (tendency towards unstable negative emotions)	• Anxious • Irritable

Adapted from: https://medium.com/psyc-406-2016/the-big-5-trait-test

The changing self

Neuroscientific research is now telling us that the brain is "plastic", which is why we now hear so much about "rewiring our brain" – it refers to "neuroplasticity". Neuroplasticity involves several changes in the brain that result from learning, including the development of synaptic connections (connecting points between brain cells), strengthening the connections, the growth of new dendrites (branched endings in brain cells or neurons), and neurogenesis (process of building new neurons in the brain). Genes are not destiny. What we do will have a profound effect on what genes are turned on or off and, therefore, our health and our experience of life.

The brain is "soft wired" by experience, and through focused interventions (like coaching) and specific projects, personality traits can intentionally change. The speed and degree of change depend on several factors, such as the person's commitment to the pursuit of change and adopting a positive mindset. A comprehensive model of personality and life outcomes is presented by the engaging Cambridge Professor and personality scientist Brian R. Little, in his internationally acclaimed book *Me, Myself, and Us*. I had the privilege of speaking with him after his guest lecture at a Sydney-based university, and he went on to reiterate how personality is moulded by many factors which go well beyond genetics.

Scientist and non-scientist personalities

Coaching science has undoubtedly become a crucible for neuroscience, psychology, medicine, and leadership knowledge, helping us identify outdated and faulty practices that have been institutionalised and continue to prevent many people from flourishing. An example of this is the misconception that, from a personality perspective (and hence neurologically speaking), there are significant differences between scientists and non-scientists. I asked this question during an interview with Dr Sarah McKay, one of Australia's well-known neuroscientists and bestselling author of *The Women's Brain Book: The neuroscience of health, hormones and happiness*. Her answer was a resounding *no* – there is *no* scientific evidence of such a difference. McKay is a (self-proclaimed) strong extrovert and, like her, there are many more STEMM professionals who are extroverts. On the other hand, what if STEMM professionals were predominantly "less extroverted"?

In 2012, results were published from an investigation of the personality traits of scientists versus non-scientists, confirming previous studies suggesting that scientists were more introverted than non-scientists, and they "prefer to be alone and are somewhat less social and less affiliative". The researchers went on to say that, since personality traits are stable from college on (their research was done on college science students), their traits are unlikely to change as the scientists enter the profession and move through a career. They therefore concluded that introverts are more likely to be attracted initially to scientific work than extroverts.

However, the paper went on to make generalisations that couldn't be substantiated by the data provided. It stated: "compared to other occupations, scientists are prone to be pessimistic, gloomy and cynical; less stable, anxious, emotionally reactive; less assertive, accommodating and more easily swayed by dominant individuals; quieter, immersed with their own thoughts, unaffiliated, with fewer connections to other people; nonconforming, independent-minded, less-rule following … and inclined to seek out variety and novel experience". Not a very nice picture, right?

Interestingly, the authors discredit the validity of their own results when describing the deficiencies in the raw data. They point out that no information was included on the employing organisations or job characteristics (for example, tenure, earnings, and scientific discipline). In addition, the study participants were all from career transition services (people who are already under psychological duress and hence in a negative state of mind). Furthermore, the study was not published in an internationally peer-reviewed journal of psychology or a related research field. Notwithstanding this, it is included in this book to introduce the concept of personality prejudices and how, over time, they may lead to exclusionist biases towards STEMM professionals.

The less extroverted personality: introverts

One of the most significant books in this area in the last 10 years would have to be the 2012 seminal book *Quiet: The power of introverts in a world that can't stop talking*, by bestselling author and ex–Wall Street lawyer Susan Cain. Reading it seven years ago was a watershed moment for me, and for many people who have been described by organisations and society at large as introverts, largely thanks to the prevalence of MBTI testing. And as we'll see, being described as an introvert in today's Western world seems to carry somewhat of an unfavourable connotation. If we recall, the Big Five model stipulates that we're on a spectrum of high or low extroversion, therefore Cain's use of the word introvert is simply an acknowledgement of a "cultural point of view", to use her words. We'll therefore use the terms "less extroverted" and "introverted" interchangeably, notwithstanding the latter's incorrectness from a psychology theory perspective.

Cain was motivated to write *Quiet* after working as a lawyer on Wall Street and observing the differences in modus operandi between herself (an introvert) and fellow (extroverted) lawyers. After extensive research and interviews with leading social and physical scientists, she concluded there was a societal dichotomy between two types of individuals that she called the "man of action" and the "man of contemplation" (here, the use of "man" denotes either man or woman). By writing about her findings and insights she hoped to create a greater balance between the two types. For the first time, there seemed to be a book that recognised attributes such as: reflective, cerebral, bookish, unassuming, sensitive, thoughtful, serious, contemplative, subtle, introspective, inner-directed, gentle, calm, modest, solitude-seeking, shy, risk-averse, thin-skinned. Moreover, she eloquently described how, in the right circumstances, these attributes can be essential to businesses, organisations, the community, and the world.

Prior to the book *Quiet*, and her subsequent *Quiet Revolution* – advocacy and educational work which has spawned dozens of books and blogs on "the introvert advantage" (one of my favourites is *The Introvert's Edge: How the quiet and shy can outsell anyone*, by Australian author Matthew Pollard) – the attributes of "man of action" seemed to dominate much of the modern leadership literature. According to Cain, these attributes are: ebullient, expansive, sociable, gregarious, excitable, dominant, assertive, active, risk-taking, thick-skinned, outer-directed, light-hearted, bold, and comfortable in the spotlight. Yes, these are very broad categories, and few individuals identify with only the man of action or the man of contemplation. However, as Cain points out, most of us recognise these types immediately because they've become embedded in our cultural stereotypes.

Interestingly, and this is what becomes a revelation to anyone who's read Cain's research in *Quiet*, to understand the whole extroversion–introversion spectrum, we must look at America's history. Prior to the early 1900s, America was a Culture of Character; that is, the ideal self was serious, disciplined, and honourable. What mattered was how one behaved in private, not so much the impression one made in public. The First Industrial Revolution then happened and corporate America was booming. The new economy called for a new kind of man: a salesman, someone with a ready smile, a masterful handshake, and the ability to get along with colleagues while also outshining them.

Dale Carnegie, the author of *Public Speaking and Influencing Men in Business*, among many other bestsellers, heeded the call and revolutionised the way business was conducted by becoming a public-speaking icon and catalysing the rise of the "extrovert ideal". Carnegie's impact on the American cultural evolution reached a tipping point at the turn of the 20th century, changing who they admire, how they should act in a job interview, what they should look for in an employee, and even what to look for in friends and romantic partners. According to Cain, historian Warren Susman described this cultural evolution as one where America shifted from a *Culture of Character to a Culture of Personality*.

Cain goes on to explain that, once Americans embraced the Culture of Personality, they became captivated by people who were bold and entertaining. "Every American was to become a performing self", according to Susman. By 1920, self-help guides had changed their focus from inner virtue to outer charm, with advice such as "to create a personality is power" and "know what to say and how to say it". An author of the times, Orison Swett Marden, who had written in 1899 the book *Character: The grandest thing in the world*, went on to publish *Masterful Personality* in 1921. The differences in advice and guidance to businesspeople are summarised in figure 20.

Figure 20: Attributes of Culture of Character vs Culture of Personality in 1920s self-help books

Culture of character	Culture of personality
Citizenship	Magnetic
Duty	Fascinating
Work	Stunning
Honour	Attractive
Reputation	Glowing
Morals	Dominant
Manners	Forceful
Integrity	Energetic

Cain's recounting of America's history goes on to make an interesting correlation: by the 1920s and 1930s, Americans had become obsessed with movie stars.

The need to appear self-assured was further embedded in people's psyche with the creation of a new concept by Viennese psychologist Alfred Adler called the "inferiority complex" (the IC, as it became known in the 1920s). On the cover of his bestselling book *Understanding Human Nature*, he asks: "Are you fainthearted? Are you submissive?", and goes on to explain how all children are born confident, however if the process of growing up is abnormal, the person can grow up to develop IC. Over time, the popular media went on to attribute being quiet, shy, introspective, or reserved to the dreaded inferiority complex. And as we've all read or seen in movies, it's even promoted – paradoxically – as a badge of honour because apparently it means that the person has also been courageous like Lincoln, Teddy Roosevelt, Edison, Einstein, and Shakespeare. Sadly, such comparisons never had the desired effect of giving hope to those "afflicted by the IC", and by the 1950s and '60s, experts were advising parents and schools to view being quiet as socially unacceptable (that is, as having a maladjusted personality) and to promote gregariousness as the ideal for both boys and girls.

University admissions officers looked for the most extroverted, not the most exceptional, candidates. In the late 1940s, Harvard's provost Paul Buck declared that Harvard should reject the "sensitive, neurotic" type and the "intellectually over-stimulated" in favour of boys of the "healthy extrovert kind". In 1950, Yale's president, Alfred Whitney Griswold, declared that the ideal Yalie was not a "beetle-browed, highly specialised intellectual, but a well-rounded man". Many deans believed that it was common sense to consider not only what the university wanted, but also what future recruiters and corporations wanted. As one dean was quoted as saying: " … they like a pretty gregarious, active type. So we find the best man is the one who's had an 80–85 average in school and plenty of extracurricular activities. We see little use for the brilliant introvert."

It seems that members of the system (that is, parents, schools, universities, and corporations) inadvertently colluded to promote the model of the modern employee: not a deep thinker but a hearty extrovert with a salesperson's personality. This even included those jobs that rarely involve dealing with the public, like a research scientist in a corporate lab. The business literature is replete with examples of STEMM individuals being described as brilliant, followed by the word "but" coupled with such words as eccentric, introvert, awkward, erratic, and so on. The popular comedy series *The Big Bang Theory* comes to mind – a show about a group of brilliant university scientists portrayed as extremely socially awkward.

Prejudice and stereotypes hinder business productivity and growth

Humans are social animals, and like other primates, we live in groups. In earlier times, the group provided considerable benefits in terms of security and protection from other individuals and groups, as well as in terms of cooperation within the group. Evolutionary theorists have proposed that genes that support cooperation and group living are selected to give individuals an advantage. Anthropologists have argued that many of the characteristics that make us uniquely human (for example, brain size and use of language) derive from our adaptation to cooperative group living in an uncertain world. Groups not only help us promote our own safety and security, they also help us to fulfil other needs, including esteem and respect. Our need to be esteemed and respected by others are aspects of a general need to belong and to feel valued by others. Another psychological term used to describe this is *relatedness*. If a person, or a group, experiences social rejection, the consequences are quite dire in terms of motivation, productivity, and wellbeing.

Relatedness is one of the three ingredients identified by psychologists Richard M. Ryan and Edward L. Deci as basic psychological needs to facilitate vitality, motivation, social integration, flourishing and wellbeing. The other two ingredients are *autonomy* and *mastery*, and they form part of their Self-Determination Theory (SDT), which is one of the fundamental concepts used in coaching psychology to develop individuals, teams and organisations to help them thrive. As such, all of our coaching programs are steeped in SDT, always endeavouring to support the client's need for autonomy, relatedness and mastery.

Perhaps people claiming not to have any untoward thoughts and feelings for the less-extroverted STEMM professional are unaware of their negative biases and subsequent harm created by such stereotypes. What happens, then, when introverted STEMM professionals encounter those stereotypes again and again, at work and in daily life? Some may be covertly singled out using seemingly harmless words like "nerd", "geek", and "bookworm". These terms may seem harmless and funny – but they're not. What happens when those stereotypes result in exclusion and rejection at work, and other social activities?

Being on the same page

This is a good time to go back to Gina, the Director of R&D. Despite her excellent performance and mostly positive feedback about her interpersonal skills, the 360-degree survey also contained very negative language in the comments section (which I later determined came from a couple of peers from Sales and Marketing). Moreover, the 360-degree results had been emailed to her by HR without a suitable debrief. As a senior leader, she had been expected by the company to be "resilient" enough to take on board the feedback and adjust her behaviour accordingly. Not surprisingly, she was feeling hurt and confused by the mixed messages, experiencing self-doubt and feeling highly stressed. She didn't know whether she could become the ideal extroverted model employee the company seemed to be demanding – at least this is what she took away from the 360-degree exercise.

Gina and I had been developing a strong coaching relationship based on trust and mutual respect. Given her recent experience, the focus of our coaching conversations was for me to empathetically listen and validate her emotions, cultivating psychological safety. Even after just one coaching session Gina reported feeling supported, less stressed, and even began interacting more positively with her peers. I appreciatively asked non-judgemental questions in all our coaching conversations to help us achieve "shared meaning" (a term used in the *New York Times* bestseller *Crucial Conversations: Tools for talking when stakes are high* by Kerry Patterson, Joseph Grenny, Ron McMillan and Al Switzler), or what is colloquially known as "being on the same page". Having a coach with a STEMM background certainly facilitated the initial creation of common ground, which is particularly important when a client has been experiencing difficulties associated with personality biases. It turns out this is much more common than many organisational leaders think.

Personality-based prejudice and legitimising myths

Personality-based prejudice, unfortunately, is nothing new and consists of many dimensions. Starting in the 1950s with an analysis in relation to the *authoritarian personality* as a result of Fascism, social sciences researchers believed that similar processes were involved in the development of a prejudiced personality. In its extreme form, the *authoritarian personality syndrome* is characterised by a simplistic cognitive style, a rigid regard for social conventions, and submission to authority figures who have a prejudice against minority groups. Another dimension to explain individual variations in prejudice is the *social dominance orientation*. Central to the theory is that the hierarchical social order is maintained through individual and institutional discrimination, often justified by "legitimising myths".

An example of a hierarchy-enhancing legitimising myth is the ideology of *meritocracy*. Meritocracy proposes that outcomes in society such as wealth, jobs, and power should be allocated according to merit rather than based on factors like gender, ethnicity or class, which are viewed as irrelevant in a cause–effect analysis. Meritocracy should increase social justice and group equality, according to those who have power, wealth, and status – who in turn often believe those who are poor have themselves to blame. They ignore the fact that people from different social backgrounds do not have the same opportunities to succeed. Hence, meritocracy can serve to legitimise social inequality and help maintain the hierarchical status quo.

Are STEMM professionals stereotyped?

The question arising from all this is: are STEMM professionals stereotyped, and do they have the same opportunities to rise to the top of organisations as non-STEMM professionals? The evidence is mixed because the STEMM community is not a homogenous group, not only in personality types but also in all the other dimensions which make humanity diverse; for example, values, culture, demographics, and physical attributes. And yet, many STEMM professionals and leaders come up against damaging stereotypes. Not unlike other groups that are currently not equality represented in positions of power (for example, women and people from other cultures), the present narrative surrounding STEMM matters needs a major overhaul.

In working with Gina, it was important to first help her focus on her personality strengths and values to increase her self-esteem, self-worth, and self-efficacy, which had all taken a battering. Gina was not depressed, however she had allowed unhelpful thoughts to distort her perceptions of others outside her team and create a negative spiral of beliefs, thoughts, emotions, and behaviours, leading to poor organisational outcomes. We used different cognitive restructuring tools to help her manage her thoughts, emotions, and environment to produce helpful behaviours to support immediate organisational goals, while also nurturing her resilience and well-being. This will be further explained in later chapters.

On the other hand, organisations need to be held accountable and be committed to fostering a positive, diverse, and inclusive work environment and culture. In Gina's and Rob's individual cases, there was a significant amount of work that needed to be undertaken by their respective organisations to ensure their environments were conducive to collaborative engagement between employees, between employees and leaders, and between leaders from different departments. To achieve this, we first worked with Gina and Rob to build their inner resources and capabilities, so they could have crucial conversations with their respective leaders, peers, and HR managers. At all times, the aim was to help both the coachee and the organisation come to a place where all parties were driven to achieve common goals, for the benefit of all concerned. Let's read on ...

CONCEPTUALISE MENTAL COMPLEXITY

*"Yesterday I was clever, so I wanted to change the world.
Today I am wise, so I am changing myself."*

Rumi

Almost every current leadership paper, article, blog, book, and website invites the reader to understand how the world is experiencing unprecedented volatility, uncertainty, complexity, and ambiguity (also known as VUCA). And as leaders move up the corporate ladder, they're faced with increasingly complex decisions. Leaders will therefore need to expand their mental complexity – their ability to embrace and make sense of uncertainty, not only for themselves, but also for others.

For leaders to be able to make balanced decisions in this complex world, an understanding of the theory of complex adaptive systems will be of great benefit. This also applies to leadership development educators, coaches, and HR leaders.

Let's take a look …

COMPLEXITY, SYSTEMS AND LEADERSHIP

Complexity is more a way of thinking about the world than a new way of working with mathematical models. Today, advances in what is known as "complexity science", combined with knowledge from the cognitive sciences, continue to inform leadership studies.

A complex system has the following characteristics:

- It involves large numbers of interacting elements.

- The interactions are nonlinear, and minor changes can produce disproportionately major consequences.

- The system is dynamic, the whole is greater than the sum of its parts, and solutions can't be imposed; rather, they arise from the circumstances. This is frequently referred to as *emergence*.

- The system has a history, and the past is integrated with the present; the elements evolve with one another and with the environment; and evolution is irreversible.

- Though a complex system may, in retrospect, appear to be ordered and predictable, hindsight does not lead to foresight because the external conditions and systems constantly change.

- Unlike in ordered systems (where the system constrains the agents), or chaotic systems (where there are no constraints), in a complex system the agents and the system constrain one another, especially over time. This means that we cannot forecast or predict what will happen.

- One of the early theories of complexity is that complex phenomena arise from simple rules. Consider the rules for the flocking behaviour of birds: fly to the centre of the flock, match speed, and avoid collision. This simple-rule theory was applied to industrial modelling and production early on, and it promised much; but it did not deliver in isolation. More recently, some thinkers and practitioners have started to argue that human complex systems are very different from those in nature and cannot be modelled in the same ways because of human unpredictability and intellect.

Consider the following ways in which humans are distinct from other animals:

— We have multiple identities and can fluidly switch between them without conscious thought. (For example, a person can be a respected member of the community while embezzling money from his company.)

— We make decisions based on past patterns of success and failure, rather than on logical, definable rules.

— We can, in certain circumstances, purposefully change the systems in which we operate to equilibrium states (think of a Six Sigma project) in order to create predictable outcomes.

Leaders need to think and act differently today

To apply the principles of complexity science in their organisations, leaders will need to think and act differently than they have in the past. This may not be easy, but it is essential in complex contexts, which is another reason executive coaching has become invaluable.

Developed by Michael Cavanagh, Professor and Deputy Director of the Coaching Psychology Unit at The University of Sydney, "The Four States of Complex Systems" illustrated in figure 21 provides a framework to facilitate coaching discussions to help leaders make effective decisions. The framework is based on the Cynefin framework created in 1999 by Dave Snowden when he worked for IBM Global Services. Used to aid decision-making, it has been described as a "sense-making device".[11]

Modified by Cavanagh to facilitate coaching conversations with leaders, the Four States of Complex Systems framework suggests that employees performing repetitive tasks require simple decision-making (the workers); managers and experts handle complicated decision-making (junior–middle management); whereas complex and chaotic environments are handled by leaders who

---

11 Snowden, D., & Boone, M. (2007). "A Leader's Framework for Decision Making". *Harvard Business Review*. Retrieved from https://hbr.org/2007/11/a-leaders-framework-for-decision-making.

possess more sophisticated decision-making capabilities (senior leadership). In other words, the higher the complexity the greater the leader's mental complexity must be. Hence, the demand for a new paradigm for leadership development.

Figure 21: The Four States of Complex Systems

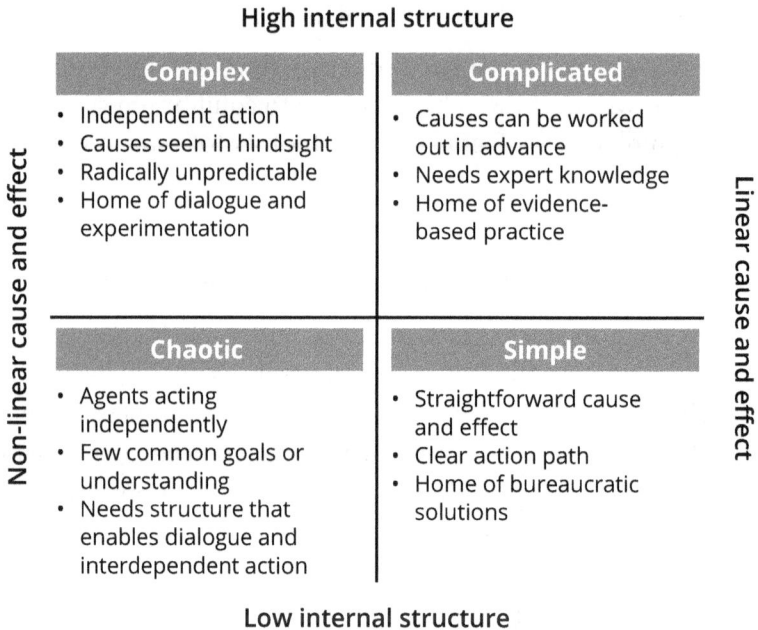

High internal structure

Complex	Complicated
• Independent action • Causes seen in hindsight • Radically unpredictable • Home of dialogue and experimentation	• Causes can be worked out in advance • Needs expert knowledge • Home of evidence-based practice
Chaotic	**Simple**
• Agents acting independently • Few common goals or understanding • Needs structure that enables dialogue and interdependent action	• Straightforward cause and effect • Clear action path • Home of bureaucratic solutions

Non-linear cause and effect (left axis) — *Linear cause and effect* (right axis)

Low internal structure

Note: Structure denotes the arrangement of and relationship between the parts.

Adapted from: Cavanagh M., "The Coaching Engagement in the 21st century: New Paradigms for Complex Times", in *Beyond Goals: Effective Strategies for Coaching and Mentoring*, David et al, 2013.

INCREASING MENTAL COMPLEXITY

First, let's define mental complexity: it is an individual's system for processing information and making sense of their environment. Research on adult development and learning suggests that increasing mental complexity is essential in helping individuals become more adaptive in complex environments, thereby providing the necessary tools to enhance performance. Robert Kegan, professor in adult learning and professional development at Harvard University's Graduate School of Education, developed a model of increasing levels of mental complexity to describe the various phases an individual might go through over time, illustrated in figure 22. This model is used by many leadership development coaches to enhance someone's ability to meet current and future challenges by developing an increasingly complex understanding of the self, others, and the systems in which the person is involved.

Figure 22: The three levels of adult mental complexity vs time

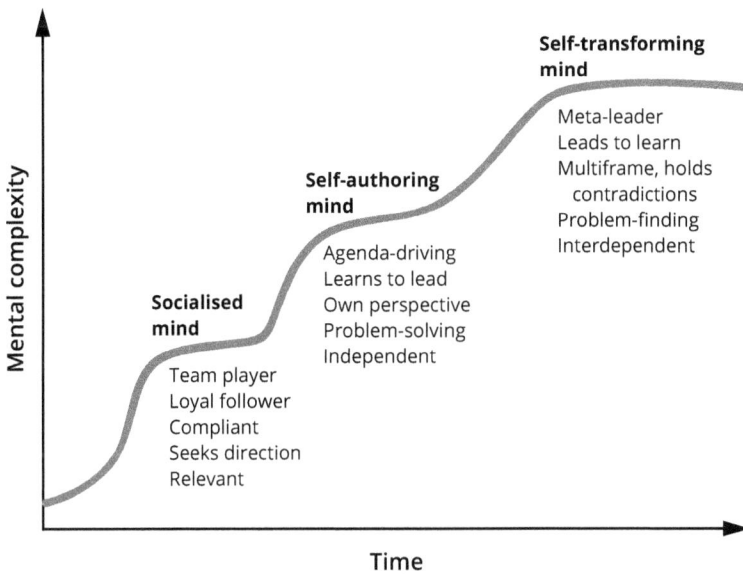

Self-transforming mind

Meta-leader
Leads to learn
Multiframe, holds
  contradictions
Problem-finding
Interdependent

Self-authoring mind

Agenda-driving
Learns to lead
Own perspective
Problem-solving
Independent

Socialised mind

Team player
Loyal follower
Compliant
Seeks direction
Relevant

Mental complexity

Time

Adapted from: Immunity to Change, Kegan & Lahey, 2009

These three adult meaning-making systems – the socialised mind, self-authoring mind, and self-transforming mind – make sense of their environment in profoundly different ways. As a consequence, they operate within their environment in profoundly different ways as well. This becomes evident in the workplace within any kind of organisational context or interaction. For instance, information flow is viewed and experienced completely differently by each adult level of mental complexity, as described in figure 23.

Figure 23: Description of the three adult plateaus

The socialised mind

We are shaped by definitions and expectations stemming from our personal environment.

This is seen mainly in our relationships with people, maintaining alignment with important others and valued "surrounds".

Highly sensitive to, and influenced by, the messages it picks up (real or imagined). The information picked up often runs beyond the explicit message. Prone to groupthink, information distortion and misunderstanding.

The self-authoring mind

We're able to step back from our social environment sufficiently to create a personal authority that evaluates and makes choices about external expectations.

We create internal coherence through alignment with own belief system, ideology and personal code. We create and regulate boundaries accordingly. We self-direct, take stands and set limits.

Places a priority on receiving the information it has sought. What is "sent" is likely a function of what is deemed others need to hear to further its agenda or mission.

The self-transforming mind

We can step back and reflect on the limits of our personal authority or own ideology. Can see that any one self-organisation or system is incomplete therefore can be friendlier toward contradiction.

Our self creates coherence through its ability not to confuse internal consistency with wholeness. It aligns with the conversation rather than with either side. It's aware that the world is in motion and that what makes sense today may not make sense tomorrow.

Advances own agenda and design while also making room for revisions or expansion. Information sending drives the agenda as well as seeks to remake the map or reset the direction. Because they welcome information, others are more likely to give it.

Adapted from: Immunity to Change, Kegan & Lahey, 2009

In a 2018 article titled "Developing Employees' Mental Complexity", authors Crane and Hartwell, from Utah State University, called for greater efforts to understand the leadership capabilities required in today's organisations, how they are developed, and how they influence important outcomes. Fortunately, many leading researchers and evidenced-based coaches are working in this important space. Michael Cavanagh (mentioned above) and Jennifer Garvey Berger (adult education expert and leadership coaching researcher) are challenging the leadership development status quo. They've produced a body of work – based on adult development and learning – which addresses the need for leaders to develop greater mental complexity due to increasing uncertainty and tension in the world.

Garvey Berger has developed coaching techniques to facilitate mental complexity transitions; that is, moving from the socialised mind to the self-authoring mind, and from the self-authoring mind to the self-transforming mind. She posits that for a coach to be able to help someone develop greater mental complexity, he/she must first assess the client's form of mind; that is, are they at the socialised mind, at the self-authoring mind, as we'll read more about next.

Figure 24: The three levels of adult mental complexity and the "self"

STAGE 3	STAGE 4	STAGE 5
The socialised mind	**The self-authoring mind**	**The self-transforming mind**
58% of adults	35% of adults	1% of adults

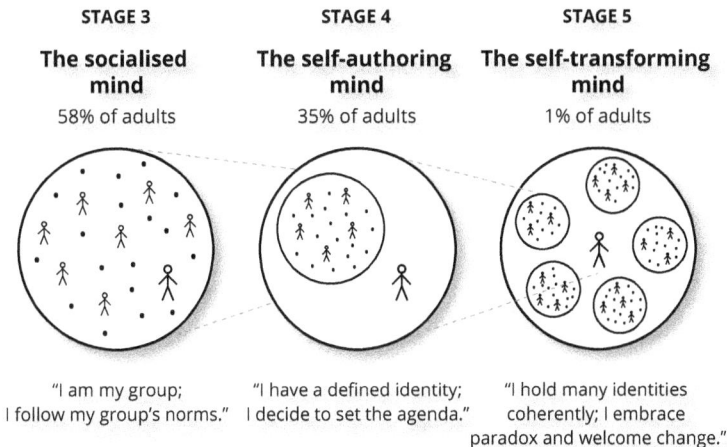

"I am my group; I follow my group's norms."	"I have a defined identity; I decide to set the agenda."	"I hold many identities coherently; I embrace paradox and welcome change."

Adapted from: *In Over Our Heads: The Mental Demands of Modern Life*, Kegan, 1994.

Assessing a leader's mental complexity

To assess a leader's level of maturity, or mental complexity, we look for key characteristics, invariably uncovered by asking carefully selected questions. In both Gina's and Rob's cases, the starting points were responsibility, conflict, perspective-taking, and assumptions about the world. As coaches, we ask ourselves when working with clients:

- What does the coachee/client take responsibility for? What doesn't he/she take responsibility for?

- What are the central conflicts in the story?

- Whose perspective can he/she take? Whose perspective is he/she stuck with?

- What assumptions about the world shape his/her view?

When we first started working with Rob he was in the socialised mind phase. Socialised mind leaders are known as "yes men" or those who need to follow company guidelines. But they can also be mavericks, such as the dot-com folks with their brilliant technology thoughts, and want to be seen as non-conformists by wearing hoodies and sitting in bean bags to do their work; or the marketing department that as a rule breaks all the rules; or the ritualistic post-modern academics; or any other group that is uniformly disloyal or uniformly different. They do not author their roles and choices.

Rob was incredibly loyal to the "behavioural rules" of his mammoth IT company (a worldwide leader); that is, he was bullish and overly aggressive, despite his own intrinsic values of love and kindness. He was living in *cognitive dissonance* (a painful psychological state resulting from behaving against one's values and beliefs), creating friction at home and frequent illness. Before coaching, this was done unconsciously; that is, Rob was repressing all manner of emotions and thoughts to cope at work. Through coaching, especially after doing a great deal of self-knowledge work, managing the cognitive dissonance became a conscious effort, causing friction in his personal relationships – where his duplicity was exposed. The psychological effort and chronic stress eventually became "too much" for his body to manage.

For a while, Rob resisted doing some of the coaching work as he couldn't take on issues he considered to be outside of "who he really was". He simply couldn't imagine a way he might change those bullish behaviours. The key aspects of our work with Rob here were to be gentle and patient as we helped him, over time, see that those externally written rules were not always helpful to him. We helped him make sense of the painful realisation that he was caught up between the voices of several important people in his life; for example, his father, his General Manager, and the much-revered company founder and Chairman. He had been a "pleaser" in the 2 × 2 Self-Awareness Model (see figure 17, chapter 5); that is, low internal awareness and high external awareness. There was much work to be done before he could become fully "aware"; that is, high in both internal and external self-awareness.

Gina, on the other hand, was further along the mental complexity spectrum. By reflecting on the questions listed above, we assessed her to be near or at the self-authoring phase of mental complexity. We spent some time helping her see her internal assumptions and how she had authored her life. Most telling was her ability to grasp with ease the differences in perspective between herself, her peers, and other stakeholders from the commercial side of the business. She was open and willing to learn about their perspectives and make sense of those views.

Given some of the differences between the STEMM worldview (for example, critical thinking, desiring certainty before making decisions, being fact-driven, slow decision-making, curiosity for the sake of it) and the commercial worldview (for example, competition for market dominance, curiosity with profitability in mind, faster decision-making for productivity, comfort with uncertainty), the sense-making and internalisation processes took some time. Only then, once it all made sense and she was able to embrace the commercial values internally, was she able to enlarge her self-authoring system. Until then, she had been an "introspector" in the self-awareness 2 × 2 Self-Awareness Model (figure 17); that is, high internal self-awareness but low external self-awareness.

Gina appreciated, wholeheartedly, understanding the history of the industrial revolutions and their impact on society's understanding of some of the less-extroverted STEMM professions. With cognitive restructuring, positivity, and mindfulness practice, she learned not to view STEMM introverts like herself as victims. To her, this was a matter of systemic intolerance of diverse personalities and cognitions. Being angry about it was unhelpful and counterproductive to her career aspirations. And as we'll see in later chapters, we coached her to find better solutions, resulting in positive outcomes for her and the company.

INVESTIGATE LEADERSHIP STYLE AND CAPABILITIES

"Leadership and learning are indispensable to each other."

John F. Kennedy

Many STEMM leaders are highly accomplished, both academically and in their careers. They've transitioned from positions where they were technical/functional experts to junior management roles, then middle management, and then to senior leadership reporting to a Managing Director or CEO. In Australia, they're likely to be described as Directors (executive), or "Head of" if they're part of a multinational with an offshore head office. If it's an Australian organisation, the roles tend to be classified as "C-suite " jobs – like Chief Technology Officer, Chief Scientific Officer, and Chief Product Officer – reporting to a locally based CEO. Sometimes head office is based in Australasia – for example, Singapore, Hong Kong, or New Zealand – although it's more likely based in one of the two major cities: Sydney or Melbourne. The point is, before reaching the top, senior leaders go through many significant changes in the company's expectations of them along the way.

Throughout the 20th century, business schools and other management educators focused on the development of skills required for linear and predictable problems. HR departments possessed "Effective Skills Management" manuals from which they would design skills and competencies matrices to train their up-and-coming leaders. They would train them in-house or through external providers in areas such as:

- Delegation skills
- Written and Verbal Communication skills
- Presentation and Public Speaking skills
- Performance Management skills
- Negotiation skills
- Interviewing skills
- Project Management skills.

However, as leaders move up the corporate ladder, they're faced with increasingly complex decisions, and will need expanded degrees of mental complexity. This is where leadership development has stalled, calling for a new approach to leadership, ideally as a science; that is, as a systematic, evidence-based process.

Leadership as an artform can be an unattractive proposition as it implies something unmeasurable and unpredictable, a quality someone is born with. Leadership as a process (or science), on the other hand, can be developed by anyone, as was discussed in chapter 2. Yes, leaders can emerge in the unlikeliest of situations, however there is always a cause and an effect. If something cannot be described, it is simply because it hasn't been studied through a scientific lens.

Unfortunately, there is no universal consensus on leadership concepts. And yet, leadership development is one of the top priorities in almost every organisation. This is undoubtedly one of the reasons for the burgeoning field of executive coaching becoming so prominent.

DO LEADERSHIP MODELS WORK?

Leadership is an extremely hot topic in business these days. With scandal after scandal being exposed in all sorts of organisations across the world, much attention has been focused on the leaders of these organisations and what they must have been doing wrong to allow their organisational culture to be corrupted.

In the academic world, according to *The Oxford Review*, in 1960 there were 2995 research papers published on leadership – and in 2017 there were 114,422 peer-reviewed papers on leadership. It's no wonder leadership development is one of the top issues keeping CEOs up at night and making HR leaders feel nervous about demonstrating a return on investment on programs they wish to run. And as anyone who has gone through the transition from being a STEMM expert to doing an MBA in the hope of developing a better understanding of organisational leadership knows, the variety of leadership styles, traits, behaviours, models, and dos and don'ts can be quite daunting and create confusion.

The confusion is exacerbated by proprietary models promoted by quasi-leadership experts and consultancies, producing easy-to-understand marketing material that is eagerly consumed by the corporate world. However, organisations are often ill-prepared to ensure that the tools they're purchasing to develop their leaders are indeed effective.

A special report in *The Oxford Review* (July 2018) stipulated that the core leadership styles most often found in the leadership literature (journals, books, business magazines, social media, and so on) are transactional, transformational, charismatic, laissez-faire, authentic, ethical and servant leadership. However, of these primary leadership constructs, only four displayed any meaningful and measurable correlations with outcomes. The others didn't provide valid research evidence as a standalone construct, nor did they result in the outcomes attributed to them.

Here are the four leadership styles that displayed meaningful and measurable correlations with outcomes:[12]

1   **Transformational leadership.** This is one of the most popular and thoroughly researched leadership styles, and as the name implies, it's a leadership process that changes and transforms people. It includes assessing followers' motives, satisfying their needs, and treating them as full human beings. Transformational leadership involves an exceptional form of influence that moves followers to accomplish more than what is usually expected of them. An all-encompassing approach, transformational leadership can describe very specific attempts to influence followers, as well as organisations and even cultures. Although the transformational leader plays a pivotal role in precipitating change, followers and leaders are inextricably bound together in the transformation process.

    Transformational leaders have a strong set of internal values and ideas, and are effective at motivating followers to act in ways that support the greater good rather than their own self-interest. They are described as possessing and displaying the transformational four "I's": Idealised influence, Inspirational motivation, Intellectual stimulation, and Individualised consideration.

2   **Authentic leadership.** This is one of the newest areas of leadership research and it's still in its formative phase. From a developmental vantage point, authentic leadership is something that can be nurtured in a leader and develops over a lifetime. Some think it is triggered by major life events or a new career. From a practical perspective, authentic leaders demonstrate five basic characteristics: a) they understand their purpose; b) they have strong values about the right thing to do; c) they establish trusting relationships with others; d) they demonstrate self-discipline and act on their values; and e) they are passionate about their mission.

---

12   These definitions are adapted from Northouse's 2016 textbook *Leadership: Theory and Practice.*

To develop a valid method to measure authentic leadership, researchers identified four components: a) self-awareness; b) internalised moral perspective; c) balanced processing; and d) relational transparency. Researchers have found there are four positive attributes that predispose a leader to developing authentic leadership: *Hope, Efficacy, Resilience* and *Optimism* (also known as HERO or Psychological Capital, which we often work with, both individually and organisationally).

3   **Servant leadership.** This is a holistic approach to leadership that engages followers in multiple dimensions (for example, relational, ethical, emotional, spiritual) so that they are empowered to grow into what they can become. It seeks first and foremost to develop followers based on a leader's altruistic and ethical orientations. When followers' wellbeing and growth are prioritised, they in turn are more engaged and effective in their work. Servant leaders see themselves as stewards of their organisations who seek to grow the resources, financial and otherwise, that have been entrusted to them. Unlike performance-oriented leadership approaches that often "sacrifice people on the altar of profit and growth"[13], servant leaders focus on sustainable performance over the long run.

The first attempt to identify the characteristics of a servant leader was made in 2002, and they were listed as: listening, empathy, healing, awareness, persuasion, conceptualisation, foresight, stewardship, foresight, commitment to the growth of the people, and building community.[14] Further research led to the identification of the following behaviours: conceptualising, emotional healing, putting followers first, helping followers grow and succeed, behaving ethically, empowering, and creating value for the community.

4   **Ethical leadership.** Some scholars argue that ethical leadership is not a unified leadership theory or model, and instead would prefer to consider the principles of leadership ethics. That is, ethics ought to be embedded in all other

---

13   From Northouse's 2016 textbook *Leadership: Theory and Practice.*
14   Ibid.

leadership theories. The word *ethics* comes from the Greek word *ethos*, which translates to *customs, conduct or character*. Ethics is concerned with the kinds of values and morals an individual or a society finds desirable or appropriate, with the virtuousness of individuals and their motives. Therefore, it's important for leaders to understand what drives them – what their values are (refer to chapter 5). In any decision-making situation, ethical issues are either implicitly or explicitly involved. The choices leaders make and how they respond in a given circumstance are informed and directed by their ethics.

Some key principles of ethical leadership have been identified and discussed in a variety of disciplines, including biomedical ethics, business ethics, counselling psychology, and leadership education. Traced back to Aristotle, these principles provide a foundation for the development of sound ethical leadership: respect, service, justice, honesty, and community.

Clearly, there is overlap between these four leadership models, and it is not uncommon for educators, consultants and organisations to create their own leadership models based on an amalgamation of these principles. Such customised leadership models usually consider organisational culture, preferred antecedents, and preferred outcomes; hence, the subjective nature of leadership development and the proliferation of concepts. The International Leadership Association periodically raises the issue of professionalising leadership, although we may still have a long way to go before anything concrete happens on that front.

Where, then, does this leave leadership development in the 21st century, and the urgent need to help STEMM leaders rise to the top?

ESSENTIAL LEADERSHIP CAPABILITIES FOR THE 21st CENTURY

Researchers and practitioners are gradually converging on the approach to leadership development, which is good news for both emerging and experienced leaders. Let's call the ability to lead "leadership capacity", which grows as the individual gains more experience, insightful reflection, and coaching interventions targeted at building a specific leadership capability. This is the approach followed by most evidence-based coaches. Not surprisingly, it is also the approach taken by the tech giant Google, which has been a champion for leadership coaching for many years. According to David Peterson, Director of Google's Center of Expertise, Leadership Development and Executive Coaching, the critical capabilities for effective leadership in VUCA times include:

- Cultivating empathy
- Managing cognition
- Fostering resilience
- Influencing others
- Deepening self-awareness.

Based on the current leadership development literature, and the latest advances in leadership coaching, we created a blueprint (refer to figure 25) to guide our coaches when working with leaders seeking to transition to more senior roles. Underneath each capability is an array of constructs which research studies suggest leadership capacity is predicated on. Therefore, each leadership program will contain customised tools, measures, questions, and exercises to ensure we are meeting our clients exactly where they are on their development journey. Examples of constructs and skills comprising the leadership capabilities are: metacognitive skills, cognitive flexibility, psychological flexibility, strategic thinking, problem-solving ability, judgement and decision-making skills, and communication skills.

The first four leadership capabilities in figure 25 (self-awareness, self-management, social awareness, and relationship management) are the four key domains comprising what is now known as emotional intelligence (EI), which will be discussed in the next chapter.

Gina and Rob each had "love of learning" in their top six VIA Character Strengths, depicted in figure 18, therefore they welcomed learning about the four most researched leadership models (transformational, authentic, servant and ethical). Furthermore, they understood how those models, in light of additional evidence from coaching psychology theory and practice, led to the leadership capabilities identified as essential in 21st century leadership, summarised in figure 8. We went on to test and agree on those capabilities needing special attention.

Figure 25: Essential leadership capabilities for the 21st century

Leadership capabilities	Improved characteristics, skills and behaviours
Self-awareness	Ability to leverage own strengths, values and drivers and mitigate behavioural derailers.
Self-management	Adaptability, positive outlook, emotional self-control and mental strength.
Social awareness	Empathy, improved listening skills, open communication skills, organisational awareness.
Relationship management and relating	Influence, coaching and mentoring, conflict management skills, team building skills, inspirational leadership (others will follow).
Sense-making	Understands and helps others understand complexity, works collaboratively, has achievement orientation and strategic thinking.
Visioning	Demonstrates values, sense of urgency, hope, community and confidence. Superior decision making skills.
Inventing	Illustrates clear priorities, creates new structures, gives permission to fail, enables others to be innovative and creative.
Leadership effectiveness	The integration of leadership capabilities results in leadership effectiveness (positive individual, team and organisational outcomes).

In both cases, as further confirmed by the results of EI psychometric tests, we decided to focus on all four competencies of emotional intelligence: self-awareness, self-management, social awareness, and relationship management. Once the EI domains were augmented (discussed in the following chapter), we went on to work on their sense-making capability through organisational/systems mapping exercises and developing sophisticated dialogue skills for effective cross-functional communication. Visioning was improved by working with different decision-making tools, whereas inventing seemed easier for Gina than for Rob, possibly due to her experience in R&D and achieving buy-in from multiple stakeholders (although improvements were still needed when dealing with commercial obstacles). Improvements were observed after six months, in both cases, and as mentioned in previous chapters, these changes translated into tangible business outcomes (for instance, improved productivity) and intangible ones (such as less stress, less conflict and greater collaboration).

However, achieving longer term and permanent behavioural changes takes time, and even more so when acquiring new leadership capabilities and expanding mental complexity (thereby building greater leadership capacity). As you'll read in future chapters, there was more work to be done. But first, let's discuss emotional intelligence (EI) in greater detail.

CHAPTER 8

THE SCIENCE OF LEADERSHIP

EMOTIONAL INTELLIGENCE ENHANCEMENT

"The greatest discovery of my generation is that human beings can alter their lives by altering their attitudes of mind."

William James

There was a time not that long ago, as recently as the late 1990s, when many people in STEMM industries were quite sceptical about emotional intelligence (EI). Those who are old enough may recall hearing, or even thinking, that EI was yet another passing fad. To them, this was just another HR course they had to do so they could simply get on with "truly important work", in the lab, the factory floor, the operating theatre, or out in the field. Fast forward to 2019 and the term emotional intelligence has become part of our daily lexicon, not only at work but also in daily life.

We now know that IQ takes second position to EI in determining outstanding leadership performance, however it's not something learnt by reading a book or attending a training course in isolation or exclusively. Developing this leadership capability (and any other, for that matter) needs to be part of a holistic and purpose-designed approach to leadership development. A leadership coaching program for STEMM professionals would be such an approach as it provides action learning and feedback opportunities about the individual's interactions with the system (the organisation, family/ friends, community, and so on) within a safe and confidential space. Furthermore, progress can be objectively measured by the coach using psychometric tools, as part of an iterative and adaptive coaching process.

If we recall our discussions in chapter 5 in relation to the extroversion personality trait (refer to figure 19) and the stereotypes and biases associated with the less extroverted (that is, the introverts), organisations may inadvertently be misjudging introverted behaviours as being less emotionally intelligent. This came up several times during the coaching conversations with Gina and how some of her R&D team members were perceived and sometimes patronised by the more extroverted colleagues from sales. It took some delving into to identify which were helpful (adaptive) introverted behaviours and which were unhelpful (maladaptive) ones such as passive aggressiveness on the part of some members of her team, and herself. She also realised that not everyone in her team was introverted; in fact, about 40% were on the high side of the extrovert personality trait spectrum. These nuanced coaching conversations helped Gina develop greater self-awareness, adaptability, organisational awareness, and conflict management skills.

DEFINING EMOTIONAL INTELLIGENCE

When we first introduce this topic in our programs, almost every client is confident that they possess excellent EI. And many do. However, as we dig deeper, we find that their understanding of being emotionally intelligent is not entirely aligned with the scientific definition, nor can they measure it. For instance, Rob was initially adamant that his EI was "above average for a senior leader in IT" because, as we jointly discovered, he thought it meant making time to listen empathetically to everyone. Indeed, in our experience, many leaders believe that EI and empathy are one and the same thing. For those who are inherently empathetic, they may even claim that EI is about indiscriminately showing all our emotions (both positive and negative). The problem here is that they're not considering the impact of those emotions on others. It's therefore essential to spend sufficient time explaining and contextualising what EI is and what it isn't.

Emotional intelligence (EI) and emotional quotient (EQ) are the capability of individuals to recognise their own emotions and those of others, discern between different feelings and label them appropriately, **use emotional information to guide thinking and behaviour**, and manage and/or adjust emotions to adapt to environments or achieve one's goal(s). There are several models currently applied by psychologists and coaches, many of which have been extensively researched. In our leadership coaching programs we often refer to Daniel Goleman's EI model because it's easy for our clients to understand how it applies to leadership. The model introduced by Goleman focuses on EI as a wide array of competencies and skills that drive leadership performance. The model outlines four main constructs or domains:

1 Self-awareness

2 Self-management

3 Social awareness

4 Relationship management.

The model then includes a set of competencies within each domain of EI, as illustrated in figure 26.

These competencies are not innate talents, but rather learned capabilities that must be worked on and can be developed to achieve outstanding performance. As such, we dedicate quite a bit of time in our coaching programs to developing these capabilities. They are discussed below very briefly, only to give the reader an idea of what each one entails. Each competency would easily warrant its own workbook. Indeed, many do exist and are explored during coaching, depending on the client's specific need.

Figure 26: Emotional intelligence domains and competencies

Self-Awareness	Self-Management	Social Awareness	Relationship Management
Emotional self-awareness	Emotional self-control	Empathy	Influence
	Adaptability		Coach and mentor
	Achievement orientation	Organisational awareness	Conflict management
			Teamwork
	Positive outlook		Inspirational leadership

Adapted from: 'Emotional Intelligence has 12 Elements. Which do you need to work on?', Goleman & Boyatzis, 2017

Self-awareness

- **Emotional self-awareness.** Leaders high in emotional self-awareness are attuned to their inner signals, recognising how their feelings affect them and their job performance. They are attuned to their guiding values and can often intuit the best course of action, seeing the big picture in a complex situation. They can be candid and authentic, able to speak openly about their emotions or with conviction about their guiding vision.

    Here, accurate self-assessment is essential, including the use of evidence-based psychometric tools and other techniques such as the 2 × 2 Self-Awareness Map discussed in chapter 5. (This serves as a good example of why training courses and coaching complement each other and shouldn't take place in isolation. The learnings from chapter 5 feed into chapter 6, and vice versa; more complex loops of learning emerge leading to higher levels of mental complexity and leadership growth.

Self-management

- **Emotional self-control.** Leaders with emotional self-control find ways to manage their disturbing emotions and impulses, and channel them in useful ways. A leader with good self-control can stay calm and clear headed in high stress or emergency situations, or remain unflappable in trying situations.

- **Transparency.** Leaders who are transparent live their values. Transparency is the authentic openness to others about one's feelings, beliefs, and actions which allows integrity. They openly admit mistakes or faults and confront unethical behaviour in others rather than turn a blind eye. To help leaders understand the interrelationships between emotions, thoughts, behaviour and environment, we use educational tools drawn from cognitive behavioural therapy (CBT), mindfulness and psychodynamics to achieve cognitive re-structuring, amelioration of intense feeling, and enhance self-regulation. These are further explained later, with additional context as they apply to our case studies. A helpful tool will be provided in chapter 10, figure 34.

- **Adaptability** Leaders who are adaptable can juggle multiple demands without losing their focus or energy and are comfortable with the ambiguities of organisational life. They are flexible in adapting to challenges, nimble in adjusting to fluid change and limber in their thinking in the face of new data or realities. An adaptable leader can meet new challenges as they arise and not be halted by sudden change, remaining comfortable with the uncertainty that leadership can bring. Cultivating this competency is closely linked to our discussions coming up in chapter 9.

- **Achievement orientation.** Leaders with strength in achievement have high personal standards that drive them to seek performance improvements both for themselves and those they lead. They are pragmatic, setting measurable but challenging goals, and can calculate risk so that their goals are worthy but attainable. They are continually learning and teaching ways to do better. They are leaders who have a sense of efficacy that they have what it takes to control their own destiny. They seize opportunities or create them rather than waiting. They do not hesitate to cut through red tape, or even bend the rules when necessary to create better possibilities for the future. Cultivating this competency is closely linked to our discussions coming up in chapter 9.

- **Positive outlook.** Positive outlook is the ability to see the positive in people, situations and events. It means persistence in pursuing goals despite setbacks and obstacles. Leaders with a positive outlook can see the opportunity in situations where others would see a setback that might be devastating to them. They expect the best from other people and believe that changes in the future could be for the better. They promote positive behaviours in their teams, which can have a ripple effect and spread across the organisation. In our coaching programs, we help our clients develop positivity by working with various models and tools drawn from positive organisational psychology, leadership and scholarship. These include tools to build *hope, efficacy, resilience* and *optimism* at individual, team and organisational levels. Some of these concepts are discussed in chapter 9.

Social awareness

- **Empathy.** Empathy means having the ability to sense others' feelings and how they see things. Leaders with empathy take an active interest in others' concerns. They pick up cues to what's being felt and thought. They sense unspoken emotions. They listen attentively to understand the other person's point of view, the terms in which they think about what's going on. As previously mentioned, many of our clients will initially come to coaching with a limited understanding of what empathy really is. This is understandable, given all the hype in social media (just do a Google search and hundreds of articles will show up, and that's just on LinkedIn).

According to Daniel Goleman, in his 2013 book *Focus: The hidden driver of excellence*, there are three kinds of empathy:

- Being able to take other people's perspective, comprehend their mental state, and at the same time manage our own emotions while we take stock of theirs is described as *cognitive empathy* ("I understand").

- In contrast, *emotional empathy* ("I feel your pain") is when we join the other person in feeling along with them; our bodies resonate in whatever joy or sorrow that person may be going through. Such attunement happens spontaneously.

- The third kind of empathy is one which leads to sympathy, and concern for others' welfare, and mobilises us to help if need be. This is *compassion* ("I'm here for you").

All three kinds of empathy are important, especially when developing organisational awareness and relationship management capabilities. It is however paramount to learn their relative applicability and appropriateness. This typically occurs with practice and coaching (either by the manager and/ or an executive coach).

- **Organisational awareness.** A leader with organisational or social awareness has the ability to read a group's emotional currents and power relationships. They can identify influencers, networks and dynamics within an organisation. Leaders who can recognise networking opportunities and read key power relationships can often be better equipped to navigate the demands of their leadership role. Complementary tools can include systems network analysis (SNA) and stakeholder management.

Relationship management

- **Influence.** Influence is a social competency. Leaders who are equipped with the emotional self-awareness and self-control to manage themselves while being adaptable, positive and empathetic can express their ideas in a way that will appeal to others. Influence is necessary for any leadership style and can be done in a way that is meaningful and effective. Leaders who are competent in influence will gather support from others and be able to lead a group that is engaged. They can mobilise the group to execute towards agreed goals.

- **Coach and mentor.** Leaders who are adept at cultivating people's abilities show a genuine interest in those they are helping along, understanding their goals, strengths, and weaknesses. Such leaders can give timely and constructive feedback and understand the power of asking the right questions to elicit solutions from their team members. Instead of telling and controlling, they teach, empower and facilitate. It's worthwhile mentioning that *coaching* and *mentoring* are viewed as two different processes in the Australian handbook *Coaching in Organisations* (2011), where:

  - *Coaching* is a collaborative endeavor between a coach and a client (individual or group) for the purpose of enhancing the life experience, skills, performance, capacities or wellbeing of the client. This is achieved through the systematic application of theory and practice to facilitate the attainment of the coachee's goals in the coachee's context.

  - *Mentoring* is a process for the informal transmission of knowledge, social capital, and the psychosocial support perceived by the recipient as relevant to work, career, or professional development. Mentoring entails informal communication between a person who is perceived to have more knowledge, experience or wisdom (the mentor) and a person who is perceived to have less (the protégé).

- **Conflict management.** Leaders who can manage conflicts effectively can help others through emotional or tense situations, tactfully bringing disagreements into the open. They help define solutions that everyone can endorse. Leaders who take time to understand different perspectives work toward finding a common ground on which everyone can agree. They acknowledge the views of all sides while redirecting the energy toward a shared ideal or an agreeable resolution.

- **Teamwork.** This is the ability to work with others towards a shared goal, participating actively, sharing responsibility and rewards, and contributing to the capability of a team. Leaders who are skilled at promoting teamwork can draw others into active commitment to the team's effort, build teamwork spirit, positive relationships and a team identity. They show others how to collaborate.

- **Inspirational leadership.** Leaders who inspire both create resonance and move people with a compelling vision or shared mission. They embody what they ask others to do and can articulate a shared mission in a way that inspires others to follow. They offer a sense of common purpose beyond the day-to-day tasks, making work exciting. These leaders can catalyse change and are able to recognise the need for change, challenge the status quo, and champion the new order. They can be strong advocates for change even in the face of opposition, making the argument for it compellingly. They find practical ways to overcome barriers to change.

LEARNING HOW TO LEAD OURSELVES

Leading in turbulent times is the new paradigm of the 21st century, placing unprecedented demands on leaders. The magnitude and speed of changes are unprecedented, all of which emphasises the need for alternative, more relevant, and adaptive approaches to leadership development. This is evidenced by the sky-rocketing levels of stress, anxiety, and related physical ailments among executives (and the population at large). (The short- and long-term impact of stress will be discussed in later chapters.) The main point is that leadership development can no longer be viewed as just learning how to lead others – equally important is learning how to lead ourselves.

The focus of effective leadership development programs is on changing leaders' mindsets and behaviours. Change is best achieved through person-centred, customised methods such as executive or leadership coaching, which help develop targeted leadership capabilities and align behaviours and mindsets with the organisational culture. This is not only promoted by organisational psychologists and coaches, it is now part of the organisational fabric of successful corporate giants like Google. However, it can often be difficult for executives to be aware of how and why they act in a certain way, which can be further exacerbated by stress.

The key missing ingredient: mindfulness training

At ProVeritas Group, we've identified that the key missing ingredient is mindfulness training. Mindfulness reduces stress and facilitates greater self-awareness. With the right professional support and practice, a person can choose more appropriate behaviours. Because of this, mindfulness is viewed by experienced executive coaches as a key enabler of effective leadership behaviours.

Unfortunately, many people are still unclear about what mindfulness is, how it contributes to overall wellbeing and, specifically, how it enhances leadership performance. In recent times, the media has questioned the effectiveness of generic training programs and tools such as apps, and whether mindfulness works for everyone. Such concerns would no doubt prompt executives, HR, and business leaders to ask, "should I include mindfulness in leadership development?"

What *is* mindfulness?

In the last two decades, the concept of mindfulness as a state, trait, process, and intervention has been successfully adopted into clinical health, psychology, and coaching contexts. Its main application is in stress management and improving emotional regulation. Mindfulness is quite a complex and ancient concept found in a wide range of spiritual practices, hence there are many definitions.

The most prominent person in bringing mindfulness into main-stream medicine and psychology is Professor Jon Kabat-Zinn. He developed the well-respected Mindfulness-Based Stress Reduction (MBSR) treatment program for stress, anxiety, and depression. And in 2012, neuroscientists D.R. Vago and D.A. Silbersweig published a framework on the physiological mechanisms of mindfulness with the view to operationalising it. They proposed a comprehensive framework to guide the next phase of major scientific breakthrough into the neurocognitive mechanisms activated during an inten-tional mental strategy like mindfulness. As a result of their 2012 study, Vago and Silbersweig described mindfulness as *a temporary state of being aware and attentive to what is taking place in the present, non-judgementally and non-reactively*. Basically, when we shift our attention inward without any judgement or interpretation, we can become aware of our thoughts, feelings, and actions. We can also shift our attention externally and become aware of what is truly happening around us. When we talk about cultivating or eliciting mindfulness, it's about facilitating the creation of an environment of awareness and attention in the present moment.

Why cultivate mindfulness?

While we are still discovering the exact physiological mechanisms engaged in mindfulness, numerous studies have demonstrated its positive impact on human wellbeing – mental and physical. This applies to both the clinical (mental or physical health issues) and the non-clinical (no mental or physical health issues) populations. Indeed, studies have shown that cultivating mindfulness helps in the development of several crucial skills for human peak perfor-mance:

- **Self-reflection and self-awareness.** R.M. Ryan and E.L. Deci stated in their 2017 book *Self-Determination Theory: Basic psychological needs in motivation, development, and wellness* that mindfulness relates to people's capacity to openly attend to current internal and external experiences. By not judging the experience and simply accepting it, we are able to calmly reflect on who we are as individuals and develop a deeper awareness of what makes us tick.

- **Self-regulation.** The first step towards an "organised mind" is to tame our negative emotions daily. Having the capacity to better understand *why* we do what we do and *why* we feel the way we feel can be incredibly empowering. This is because once we are aware of unhelpful thoughts and emotions, we can consciously choose a more helpful and positive course of action. In other words, we can appraise and therefore self-regulate adaptively.

- **Positive relationships.** When we are aware of how our thoughts and emotions drive our behaviours, and can reflect, consider, and adopt more helpful or positive thoughts, our behaviours will also become more helpful and positive. This will in turn have a positive impact on our personal and work relationships. As people notice less reactive and calmer behaviours in us, the quality of our relationships will improve. Our social network will inevitably expand in a more meaningful way, which leads to greater emotional wellbeing.

- **Competence.** Mindfulness allows people greater insight, self-awareness, and the self-reflection necessary to ensure that their behaviours are congruent with their values. With improved self-reflection and self-awareness, they can identify the most effective pathways to achieve their goals, leading to improved confidence and competence.

- **Stress reduction.** Mindfulness can reduce stress and its damaging effects on the body, therefore it has sparked a growing interest in its influence on the immune system and disease risk. There is evidence that mindfulness may dampen the activity of genes associated with inflammation – essentially reversing molecular damage caused by stress. It is early days, however the research looks promising so far.

It's easy to recognise how improvement in all these areas will improve a person's performance at work. Being less reactive, developing better relationships and reducing stress will benefit everybody in the workplace.

We are no longer looking at leading "from turbulence to stability". Leading in turbulent times is the new paradigm for businesses.[15] Even the most experienced leaders are finding it increasingly challenging to perform sustainably in today's VUCA (volatile, uncertain, complex, and ambiguous) workplace.

Research suggests that 30% to 50% of senior leaders are pushed out, fail or quit within 18 months of a new appointment. Reasons for new appointments include divisional or international transfers, changes from mergers and acquisitions, internal promotions, and external hires. While the reasons are unique to each individual's circumstances, the underlying themes are mostly linked to working in today's VUCA environment. This was certainly implied by studies from The University of Sydney showing that a large percentage of executives seeking coaching are experiencing high levels of stress and anxiety.

It has long been established that high levels of stress and anxiety are detrimental to individual and organisational performance and wellbeing. As such, businesses need to find and apply interventions that seek to effectively reduce stress and thereby improve performance.

Being mindful is the opposite of being in "automatic" or "auto-pilot" mode, not knowing why we have acted in a certain way and not knowing why we feel the way we do in certain situations. Cultivating mindfulness through coaching has been shown to improve performance by contributing to several areas:

- **Listening skills.** The leadership coach facilitates a calm and curious focus of attention on a client's state of wellbeing throughout the coaching program. This improves the listening skills of the client.

- **Less reactive.** Mindfulness helps clients become less reactive to ups and downs of change, be more curious and objective, and be more self-aware and self-reflective about what's

---

15   As poignantly warned by Jorrit Volkers, Dean of Deloitte University (EMEA) and
      Conference Chair at the 19th International Leadership Association's Global Conference
      2017.

working and what's not. Clients develop the mental space to be proactive.

- **Goal achievement.** Coaches help bring mindful attention to the client's motives and goals, ensuring alignment with the client's values, thereby increasing motivation and the likelihood of goal achievement.

- **Cognitive and emotional agility.** Once the client's self-awareness and self-reflection have been expanded, the coach can then foster cognitive and emotional agility.

- **Emotional intelligence.** Once cognitive and emotional agility are developed, the coach can facilitate nuanced interventions to help the client manage his/her emotions, and the emotions of others. That is, emotional intelligence is enhanced.

- **Sense-making capability.** Cultivating mindfulness also assists clients in making sense of experiences and connect the dots; that is, it aids in building sense-making capacity – a key leadership capability.

- **Perspective-taking capacity.** In addition, when a client can observe a situation dispassionately, multiple perspectives become evident with greater ease, enabling him/her to make more balanced decisions.

DEVELOPING EMOTIONAL INTELLIGENCE IN ALL LEADERS

All leaders need to develop EI, not just STEMM leaders, as evidenced by the increased focus and abundance of courses at present. The workplace will continue to change dramatically with the replacement of humans with AI, in manual and technical jobs. We'll therefore see the proliferation of jobs requiring individuals, and leaders, with sophisticated people skills to collaborate on the generation and implementation of creative solutions, often across multiple demographics. EI will continue to rise as a key competency – one that differentiates between good leaders and remarkable ones.

For Rob, developing EI was relatively easy since he had always worked in sales. As soon as he completed a Bachelor of Information Systems Engineering in his early twenties, he accepted a technical sales role with a large software multinational. He seemed comfortable speaking about his area of expertise, however he had some blind spots concerning some behavioural patterns that were adversely affecting his relationships. These blind spots stemmed from feelings of social inadequacies, projected as somewhat aggressive and competitive behaviour. We used several techniques to help monitor his thoughts and emotions, such as reflective journalling (with guided questions). As we'll see in the following chapter, he learnt to notice and process unhelpful thoughts and emotions (we called that "internal churn") more adaptively using a combination of coaching tools. We also implemented mental rehearsals before important meetings, helping him improve his external and organisational awareness, and his ability to relate to others. Within a few months he was inspiring and influencing his peers as well as members of his team, achieving unimaginable sales in less than 12 months. Plus there was talk of a future promotion.

In Gina's case, the process of bolstering each EI domain was somewhat longer. Gina had spent most of her career to date in a technical environment dealing predominantly with scientists both in academia (during her PhD and part-time lecturing) and in industry. She had developed enough emotional intelligence to successfully manage a stellar career as a scientific leader, however it was now time to expand her self-awareness and social awareness even further. We focused on learning more about the complexities of the commercial pharmaceutical world for her to communicate more effectively with her Sales and Marketing colleagues. This entailed undertaking post-graduate business courses, which her employer was happy to sponsor. Our coaching sessions continued once a month to keep working on her interactions at work, providing her "emotional scaffolding" to take on new projects that exposed her to commercial decision-making opportunities. As she learnt more about systems theory and market forces, she developed greater empathy for her colleagues while also honing her organisational awareness (and savviness) to push back when warranted. Of course, this was quite taxing at times since it was a significant departure from her original technical management training. We leveraged her values – curiosity, leadership, bravery, love of learning, hope and kindness – to stay motivated and positive as she gradually strengthened her work relationships outside of her R&D department. Nurturing her resilience and wellbeing became important goals in order to achieve sustainable work performance.

CHAPTER 9

NURTURE RESILIENCE AND WELLBEING

*"Wellness is the complete integration of body, mind, and spirit –
the realisation that everything we do, think, feel, and believe has
an effect on our state of well-being."*

Greg Anderson

Interpersonal neurobiology, social neuroscience, affective neuroscience and sociophysiology are among the emerging fields of study attempting to bridge the gap between the biological and social sciences, as explained by American clinical psychologist and professor of psychology at Pepperdine University Dr Louis J. Cozolino, in his book *The Neuroscience of Human Relationships: Attachment and the developing social brain*, 2nd Edition. Even though it's now believed by scientists that personality is 50% a result of genetics, interpersonal neurobiology assumes that the brain is a social organ moulded by experience. Through interdisciplinary exploration, it seeks to discover the workings of experience-dependent plasticity. At the core of interpersonal neurobiology is a focus on the neural systems that organise attachment, emotion, attunement, and social communication. And as we're about to see, this is directly relevant to organisational wellbeing and leadership.

THE BIOLOGY OF STRESS

Understanding what goes on in our bodies when we experience stress helps us develop a deeper appreciation of the devastating consequences when stress is chronically high. Stressors are events that trigger reactions – that is, stress responses – which are physiological and emotional responses to stressors. The autonomic nervous system (ANS) and the endocrine system produce physical reactions to stressors. The ANS consists of the parasympathetic system (relaxed, or rest-and-digest state) and the sympathetic system (the fight-or-flight reaction). The endocrine system releases hormones (cortisol, epinephrine, norepinephrine) and these hormones in turn trigger physiological changes.

Common emotional responses such as anxiety, fear, and depression can often be moderated or controlled through the somatic nervous system (SNS), which is entirely under our control. However, our responses can be either effective or ineffective, and are influenced by factors like personality, health, life experiences, and our coping skills. For instance, individuals who are high on the neuroticism spectrum in the Big Five model (see chapter 5) tend to react more strongly to stressors and may have difficulty coping adaptively. On the other hand, people who are high in openness may easily view stressors as challenges and learning opportunities. It's also important to note that past experiences can influence the perception of a potential stressor. In all these instances, leadership training or coaching can help a person develop adaptive coping mechanisms.

Figure 27 illustrates the ANS sympathetic fight-or-flight response experienced by the body when a person perceives a potential threat. Once a threat is detected the body responds automatically, and this may happen for good reason; for instance, if you're about to be mugged. However, these reactions can become problematic and even harmful when undergoing a major life change (for example, moving house or the death of a partner), or if stress levels are persistently and chronically high; for instance, if the person is working in a highly toxic workplace or is experiencing bullying and believes they have no control over the situation.

Figure 27: The fight-or-flight threat response (sympathetic system)

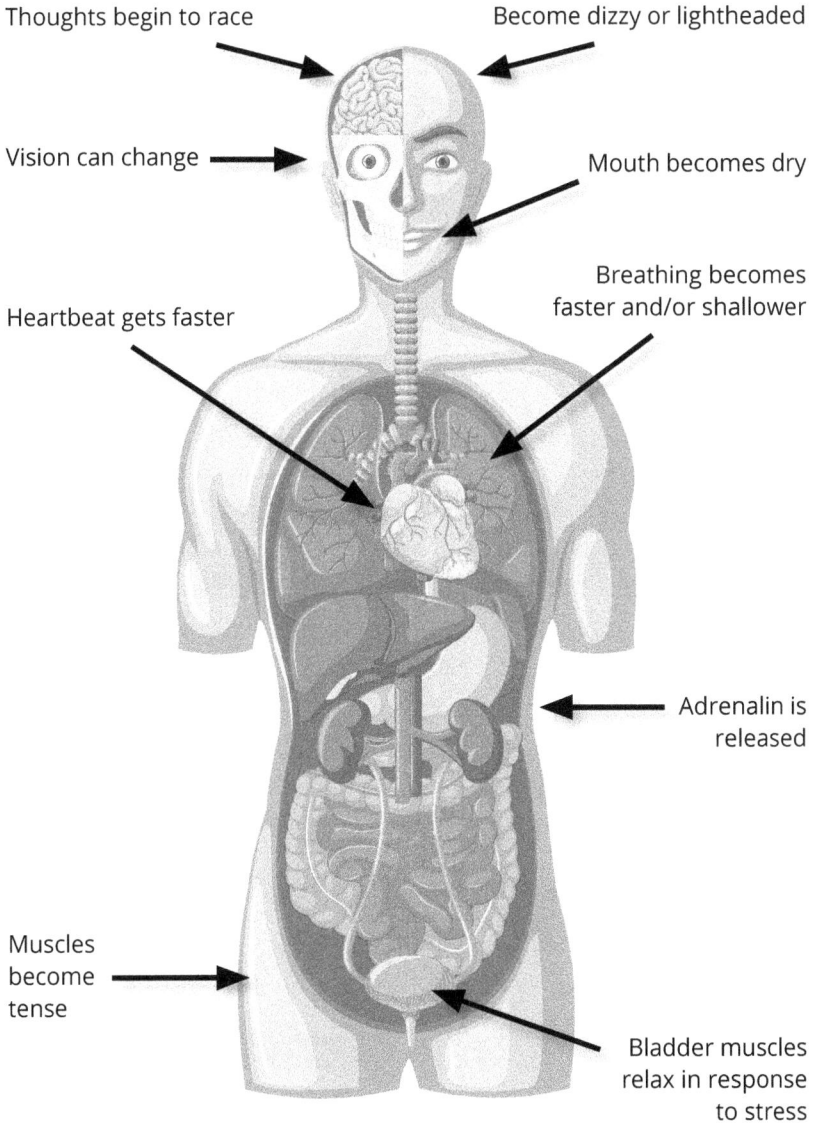

Thoughts begin to race

Become dizzy or lightheaded

Vision can change

Mouth becomes dry

Breathing becomes faster and/or shallower

Heartbeat gets faster

Adrenalin is released

Muscles become tense

Bladder muscles relax in response to stress

Image: Shutterstock.com.
Adapted from: Psychologytools.org.

Burnout – a growing workplace problem

The topic of wellbeing at work has been gaining momentum around the world for quite some time now. Indeed, a new policy document has been published by the Australian Government to assist organisations to proactively achieve wellbeing in the workplace for their employees. There is an abundant amount of information currently available on wellness and wellbeing programs, offered by medical professionals, psychologists, and coaches. Many large organisations offer such programs freely to their employees, including one-on-one coaching. (I have been involved with such programs for large Australian banks.)

But employee burnout continues to escalate at alarming rates, with an estimated $125 billion to $190 billion in related healthcare spending each year, and burnout has also been linked to the development of type 2 diabetes, coronary heart disease, gastrointestinal issues, high cholesterol, and even death for those under the age of 45. Indeed, burnout is included in the World Health Organization's eleventh revision of the *International Classification of Diseases* as an occupational phenomenon. It's described in the chapter "Factors influencing health status or contact with health services", which includes reasons for which people contact health services but that are not classed as illnesses or health conditions. Burnout is defined by the WHO as "a syndrome conceptualized as resulting from chronic workplace stress that has not been successfully managed. It is characterized by three dimensions:

- Feelings of energy depletion or exhaustion;

- increased mental distance from one's job, or feelings of negativism or cynicism related to one's job; and

- reduced professional efficacy."

Understanding what goes on in our bodies when under extreme and sustained stress (just like many leaders now experience within today's VUCA workplace) helps us to develop a deeper appreciation of the devastating consequences. More importantly, it informs us about the best measures to be taken to restore afflicted people, and to hopefully avoid getting there in the first place.

Richard Boyatzis – Professor of Organizational Behaviour, Psychology, and Cognitive Science at Case Western Reserve University – has brought together the intricate scientific concepts relating to poor leadership, toxic workplace environments, neuroscience, and wellbeing, and their relationships to stress.[16] I have summarised these as follows:

- Chronic stress causes and perpetuates a cycle of cognitive, perceptual, and emotional impairment via the activation of the sympathetic nervous system, generating harmful doses of hormones such as epinephrine, norepinephrine, and corticosteroids.

- At sustained high levels, these hormones can increase blood pressure, large muscles prepare to fight or flight, the brain shuts down non-essential neural circuits making the individual less open, flexible and creative, and the healthy function of the immune system is reduced. This cycle is commonly known as the "sacrifice syndrome".

- On the other hand, engaging the parasympathetic nervous system has the effect of reversing (in most cases) the effects of chronic stress by releasing oxytocin and vasopressin, which are pituitary neuropeptides shown to affect social processes. Their production will activate neural circuits, decrease blood pressure, improve the immune system, and promote positive behaviours such as wanting to understand another person, and feeling hopeful, at peace, optimistic, and excited about the future. This is known as the "renewal cycle".

- An expert trainer or coach will include interventions that promote mindfulness, always with compassion and eliciting hope in the leader. As such, effective leaders will inspire others by creating and maintaining resonance.

---

16   Keynote speech at the 2018 Coaching in Leadership and Healthcare Conference at Harvard McClean Medical School IOC, Boston.

Research is ongoing, and it's exciting to keep a watchful eye on these advances, especially as they will inform us on the most effective interventions to help leaders avoid burnout. This research can also provide the scientific foundation for what we already know about great leaders, as wisely articulated by Richard Boyatzis[17] (and adapted by the author):

- Great leaders inspire through vision and hope.

- They spread compassion.

- They are mindful, and attuned to their emotions, thoughts, behaviours, body, and environment.

- Great leaders inspire others by creating and maintaining resonant relationships.

MANAGING STRESS

Many organisational leaders remain unaware of the fact that not every individual can be managed or led in the same way. Not everyone thrives under sustained high levels of pressure (does anyone?) or under the same workplace conditions. The perfect example is the open office design craze that swept the corporate world a few decades ago, leading to high levels of stress for people who are less extroverted and prefer quiet environments. Similarly, excessive numbers of meetings, long work hours, and "stretch goals" have become the new normal. No longer are people allowed to reflect, think, and create individually at work. If it's not done in a meeting, among a group of people, it doesn't seem to be viewed as work. This is having dire consequences in terms of individual health and wellbeing, as reported in many business and scientific magazines, blogs, and whitepapers.

It all begins in primary school and is observed right through high school, college or university, and finally the workplace – the needs of the approximately 50% of the population who are introverts

---

17 2018 Coaching in Leadership and Healthcare Conference at Harvard McClean Medical School IOC, Boston.

are ignored. Nobody will get the best out of introverts in a brainstorming session, whether at university or at work. Susan Cain, the author of *Quiet* and "patron of introverts", has made these issues part of the modern organisational narrative, especially in her native America. Cain has presented compelling data on the detrimental health impacts of imposing "extroverted conditions" on introverts.[18] She is on a quest to influence education and business leaders in embracing a more inclusive approach to designing schools, offices, teaching practices, and workflow practices.

The Quiet Revolution website was discussed with Gina during her coaching sessions, which she found incredibly validating and empowering. In our discussions, we were able to unpack how her behaviour would change during meetings and other interpersonal interactions with some of her more extroverted colleagues. At times, she would unconsciously feel under threat and react in an uncharacteristically blunt manner. Some of her behaviours were negative – for example, becoming agitated or aggressive – and we worked on restructuring her thoughts (she had a tendency of unfairly "labelling" the more gregarious sales and marketing folk as "showoffs").

After several coaching interventions, she came to appreciate and embrace differences in personality and manage her physiological reactions whenever she felt threatened. She also downloaded helpful meeting templates, suggestions, and educational material about how to structure "mixed-temperament" team meetings, which she discussed with HR, her team, and her commercial peers. Most leaders in the business found the "inclusive meetings checklist" to be quite enlightening, and it became part of the company's standard operating procedure for managing meetings.

Other stress-management tools discussed during this coaching included social support, regular exercise, good nutrition, time management skills, clear communication, and relaxation techniques (for example, visualisation and deep breathing). Among the cognitive restructuring techniques, we discussed modifying expectations,

---

18   Keynote speech at the 2018 Coaching in Leadership and Healthcare Conference at Harvard McClean Medical School IOC, Boston.

monitoring self-talk, living in the present, being flexible, and laughter.

IT'S ALL IN THE MIND

There is no right or wrong personality type. One of the most important objectives in executive coaching is learning **metacognitive skills**, which is learning how to think about thinking. In cognitive restructuring, we show our clients how to manage their thoughts adaptively. If we go a step further, we can learn to apply a coaching model that Professor Anthony Grant named the "house of change", which is about optimising our thoughts, emotions, behaviours, and situations or environments in order to achieve our goals.

Figure 28: The "house of change" model

My goal is ...

Situation	Behaviour

Thoughts	Emotions

Adapted from: Grant & Greene, 2004

These four factors are interrelated, and by changing one, the others will also change. Think of the difference it can make to an introvert if the environment is a calm one. He/she will be able to think up better and more creative solutions to a problem, feel proud and engaged at work, and design or invent the solution to a wicked problem. Now that would be an extraordinary achievement, right? What, then, is the difference between the extrovert's and the introvert's brain from a biological perspective? The answer has to do with the way the brains are wired, and unless extroverts understand this, they may get the mistaken idea that introverts are antisocial or rude.

One major difference between the brains of introverts and extroverts is the way they respond to the neurotransmitter dopamine. Dopamine is a chemical released in the brain that provides the motivation to seek external rewards like earning money, attracting a romantic partner, or getting selected for a high-profile project at work. When dopamine floods the brain, both introverts and extroverts become more talkative, alert to their environment, and motivated to take risks and explore. Both introverts and extroverts have the same amount of dopamine available. However, according to Scott Barry Kaufman, the Scientific Director of The Imagination Institute in the Positive Psychology Center at the University of Pennsylvania, the difference is in the activity of the dopamine reward network. It is more active in the extroverts' brains that in the introverts' brains. At a noisy and crowded rock concert, extroverts become more energised than introverts. They buzz with an enthusiastic rush of positive feelings, while the introverts feel overstimulated.

Introverts, on the other hand, use more of a different neurotransmitter, called acetylcholine, according to author Christine Fonseca in her book *Quiet Kids: Help your introverted child succeed in an extroverted world*. Like dopamine, acetylcholine is linked to pleasure. However, for introverts the feeling of pleasure is heightened when they turn inward. Furthermore, it seems to power the abilities to think deeply, reflect, and focus intensely on just one thing for a long period of time. It also helps explain why introverts like calm environments, since it's much easier to turn inward when they're not attending to external stimulation. So, when an introvert is relaxing at home in quiet solitude, lost in a book or watching TV, they're basking in the enjoyable effects of acetylcholine.

Acetylcholine is also linked to the parasympathetic (relaxed state) part of the autonomic nervous system, previously mentioned. When we engage the parasympathetic system, our body conserves energy and we withdraw from the outer environment. Our muscles relax, energy is stored, food is metabolised, pupils constrict to limit incoming light, and our heart rate and blood pressure decrease. In other words, the body gets ready for contemplation, which introverts seem to like the most. When the sympathetic system is activated, thinking is reduced, and we become prone to making snap decisions.

Hence, it's fair to say that the differences between introverts and extroverts are indeed in the mind, or more accurately, in the brain. To try to change an introvert to become an extrovert might be incredibly difficult, if not impossible. Conversely, asking an extrovert to become more introverted may also be mission impossible. In the interests of everyone's health and wellbeing, it is best to accept our differences, embrace diversity, and work, play, and live in harmony.

BACK TO MINDFULNESS

We spoke about mindfulness in chapter 8, however since it's such a pivotal aspect of any leadership development initiative in today's VUCA world, it deserves another look. Many large organisations have already made mindfulness practice training an integral part of their internal leadership programs. Sadly, the majority still don't, possibly because their HR leaders are grappling with other major transformational shifts, such as digitisation of HR systems and talent retention challenges in a globalised workforce.

Internal and external executive coaches have identified the need to make the links between mindfulness, neuroscience, leadership effectiveness and organisational performance more explicitly articulated. As more multidisciplinary scientific research emerges and practices are validated, mindfulness will likely become part of mainstream leadership development programs in business schools and executive coaching programs on multiple platforms.

One such program is the eight-session "white label" course from called Mindfulness X, where the focus is on developing the client's *mindful attention* (depicted in figure 29). Being mindful is the opposite of being in automatic or in auto-pilot mode, not knowing why we have acted in a certain way and not knowing why we feel the way we do in certain situations.

Figure 29: How mindfulness works

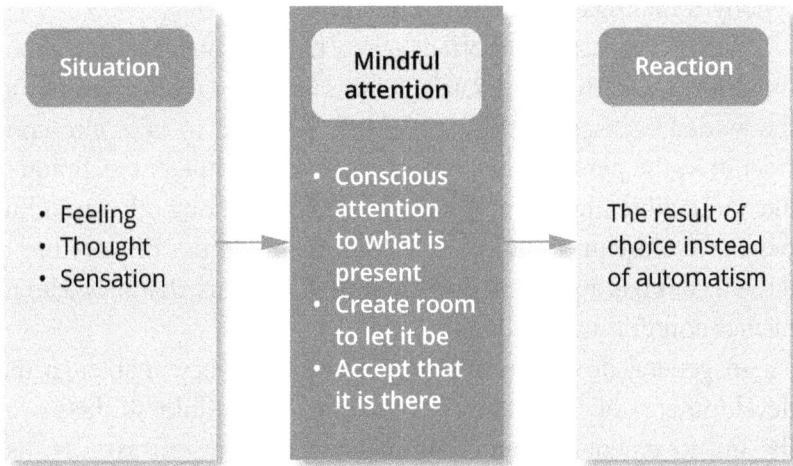

Situation	Mindful attention	Reaction
• Feeling • Thought • Sensation	• Conscious attention to what is present • Create room to let it be • Accept that it is there	The result of choice instead of automatism

Adapted from: Mindfulness X course material from Positive Psychology
https://positivepsychology.com/course/mindfulnessx/

Not surprisingly, combining mindfulness (which gives us a choice point to think of better options) and cognitive restructuring (which shows us which thoughts are helpful and which ones aren't) can be one of the most powerful techniques anyone can master. Known as MBCT (mindfulness-based cognitive behavioural therapy) it was initially developed for stress management purposes. Nowadays it's widely researched, accepted and used in multiple settings. As such, operationalising it can turn it into an easily recalled mental program. In turn, this mental program could become a formidable building block in the development of other complex and essential capacities for high leadership performance. We'll come back to this in chapter 10. Now we must discuss the closely related construct of resilience ... and beyond.

It is very important for mindfulness to be introduced as a secular practice based on scientific research. Unfortunately, many mindfulness practitioners, aided by mainstream media, have created the image of someone in the lotus position as representing mindfulness. An unconscious negative bias against meditation could hinder a client's willingness to engage in mindfulness, let alone build a habit. Mindfulness is not a religious or spiritual practice, and it is not a synonym for Yoga or other prescribed meditation.

This is especially important when having conversations about mindfulness in the corporate environment, and even more so in the evidence-based world of STEMM. We need to take into consideration the person's preferences and explore different techniques like guided breathing and guided journaling, among other mindful practices. More importantly, mindfulness should not be introduced if the leader has a physical or mental health history that may lead to his/her condition worsening.

In general, however, mindfulness can be a key enabler in the development of leadership capabilities. Mindfulness helps in the improvement of numerous mental and physical aspects that lead to peak performance, such as self-awareness, self-reflection, self-regulation, positive relationships, and improved competence. When skilfully incorporated into leadership development programs, leadership performance will increase as a result of improved listening skills, cognitive and emotional agility, goal achievement, emotional intelligence, sense-making, and perspective-taking capacity.

Rob – the Sales Director in the IT multinational – struggled to embrace mindfulness for a long time because he associated it with meditation, which he thought was too "new age" for his liking. He also thought it was "too feminine" and that his work and personal male friends would laugh if they were to find out. Yes, we gently unpacked a great deal in relation to his outdated gender stereotypes and how there was no room for those views in current leadership practice, let alone the future. With a socialised mind backdrop, wording was carefully chosen so as not to incite sudden opposition. Finding respected role models helped.

Back to mindfulness though; when explained the concept from a scientific perspective, taking the client through a specially designed mindfulness course for STEMM leaders, we've found that clients embrace it more readily. Indeed, Rob found it much easier to conceptualise. By trialling different types of mindful exercises that easily fit into his lifestyle, he became more open and gradually incorporated this into his daily routine. Eventually, he learnt to "stop–breathe–think". We called this the SBT model, which can be found in figure 34. Rob became less reactive at work and at home. Mindfulness techniques were incorporated into the "breathe" step, which helped him slow down his mind to then think things through calmly and rationally (using cognitive restructuring techniques). His mind and body thanked him for it as he learned to manage stress more adaptively and make better decisions.

This proved to be a game changer for Rob given the high-pressure and fast-paced nature of sales senior leadership in technology. Within three months he reported being able to manage his emotions more effectively, feeling less stressed and being able to have calmer conversations with colleagues and clients. After six months of coaching he was exceeding sales targets. Within 12 months of coaching Rob had become the star performer nationally after designing innovative commercial partnerships and delivering multimillion-dollar deals involving several technology partners, government and academia. Rob was well and truly on his way to reaching new heights in his career, as well as in his personal life.

But everything was about to change for Rob. He was about to become a father and had just purchased his first home with his wife (who was also moving up the corporate ladder in a different field). And while he was a "rock star" in one of the world's largest IT companies, he had grown increasingly restless from being in sales for close to 14 years. Because of his engineering background he had a natural ability to understanding how different parts of the organisation, including people, were interconnected to yield business outcomes. He was eager to lead a whole company. We continued to work together, focusing on developing the necessary internal resources to prepare him for the (significant) life changes he was about to embark on. We therefore turned our attention to nurturing resilience and wellbeing ...

BEYOND RESILIENCE

Resilience is a complex construct that may have specific meaning for a particular individual, family, organisation, society, and culture. Individuals may be more resilient in some domains of their life than others, and during some phases of their life compared with other phases.

The topic of resilience was discussed from a comprehensive, interdisciplinary perspective during the 29th Annual Meeting of the International Society for Traumatic Stress Studies, held in Philadelphia, Pennsylvania, in November 2013. The discussion was chaired by Steven Southwick, M.D. The outcomes of those discussions were captured in the article "Resilience Definitions, Theory, and Challenges: Interdisciplinary Perspectives", published by the *European Journal of Psychotraumatology* in 2014. The following definition emerged to be the most fitting: **"resilience is associated with the ability to employ a variety of coping strategies in a flexible manner depending on the specific challenge, and then to use corrective feedback to adjust those strategies"**. Due to its complexity, experts agree that it isn't yet possible to measure resilience, despite the biometric technologies and computational power currently available. No doubt we will be able to soon.

Notwithstanding the attention given by the media and corporate trainers to this psychological construct, we need to look **beyond resilience** in order to thrive in the 21st century. To undertand why this is the case, we first need to talk about another construct, known as "mental toughness". Mental toughness (MT) has been defined as "a plastic personality trait which determines how individuals respond mentally to stress, pressure, opportunity and challenge, irrespective of prevailing situation".[19] MT comprises four elements, described

---

19   Clough & Strycharczyk, 2011.

in figure 30. These can be readily measured using psychometric tools, and therefore monitored as they grow throughout a coaching program. In terms of psychology theory, resilience comprises *control* and *commitment*, which enables us to survive. Mental toughness goes beyond that since it comprises *control*, *commitment*, *challenge* and *confidence*, which enable us to thrive. Studies have linked MT to high performance, adoption of positive behaviours, and wellbeing in multiple contexts (such as leadership, the workplace, sports and the arts). It's important to note that mental toughness does not mean being hard, inflexible, harsh or unemotional. It simply means being mentally strong.

Figure 30: The four dimensions (4Cs) of mental toughness

Scale	What this means – how an MT person thinks ...
Control	I really believe I can do it.
	I can keep my emotions in check when doing it.
Commitment	I promise to do it – I'll set a goal.
	I'll do what it takes to deliver it (hard work).
Challenge	I am driven to do it – I will take a chance.
	Setbacks make me stronger.
Confidence	I believe I have the ability to do it.
	I can stand my ground if I need to.

WHEN THE GOING GETS TOUGH ...

Depending on your age, you'll probably be able to finish that line in your head – *the tough get going* – from the 1985 Billy Ocean hit song. Both Gina and Rob learnt the true meaning of it as we worked on each of the four MT dimensions from the table above. As might be expected, given that both are high achievers, commitment wasn't much of a problem, and neither was challenge. Both Gina and Rob were on the high side of the conscientiousness spectrum in the

Big Five personality traits model discussed in chapter 5 (refer to figure 19).

Our emphasis was on bolstering control and confidence in both cases, although within different contexts and due to different root causes. When discussing control in greater detail with Gina, we considered another personality trait – neuroticism – for which she was slightly on the higher side. This information was valuable as it helped us focus on self-regulation techniques such as cognitive restructuring, visualisation and mindfulness. The SBT tool described in the next chapter proved to be very helpful in strengthening her control and confidence. Gina took her R&D team to new heights of achievement, establishing a global position as a centre of innovation and helping the company achieve 20% market share growth in a key, yet highly crowded market sector in 12 months. All this while she enjoyed better health and stronger personal relationships (she married a kind-hearted and highly accomplished healthcare executive).

Rob mentioned the *impostor syndrome* (persistent internalised fear of being exposed as a "fraud") and how it eroded his self-confidence. We tapped into his innate curiosity and problem-solving skills and viewed it as another challenge that could be solved systematically. They enhanced their *agency* (capacity to act independently in an environment) by articulating their numerous achievements to date. They each identified new goals and listed the resources and action items needed in as much detail as possible. The development of SMART goals is discussed further on. We also applied the SBT tool to challenge the *impostor syndrome* thought whenever it popped into their heads. These exercises were captured in their regular reflective journalling, and over time their confidence radically improved, which was reflected in explicit acts of inspirational and servant leadership. Rob maintained high levels of wellbeing as his family grew in number and he accepted a more senior role as general manager of a business unit, all within 18 months.

HIGH PERFORMANCE IN A VUCA WORLD

There's a whole field of coaching science dedicated to optimising and maintaining high performance in sports athletes and in the performing arts. Many aspects of this field are often applied by executive coaches when designing leadership programs, especially for senior leaders. Much of it is either discussed or touched on in this book, which is largely focusing on expanding emotional strength and capacity; and mental strength and capacity. The other areas are a) physical strength and capacity – without which it's very difficult to build the other areas – and b) self-actualisation or spiritual capacity (the choice of term is an individual choice).

Figure 31: The high-performance pyramid

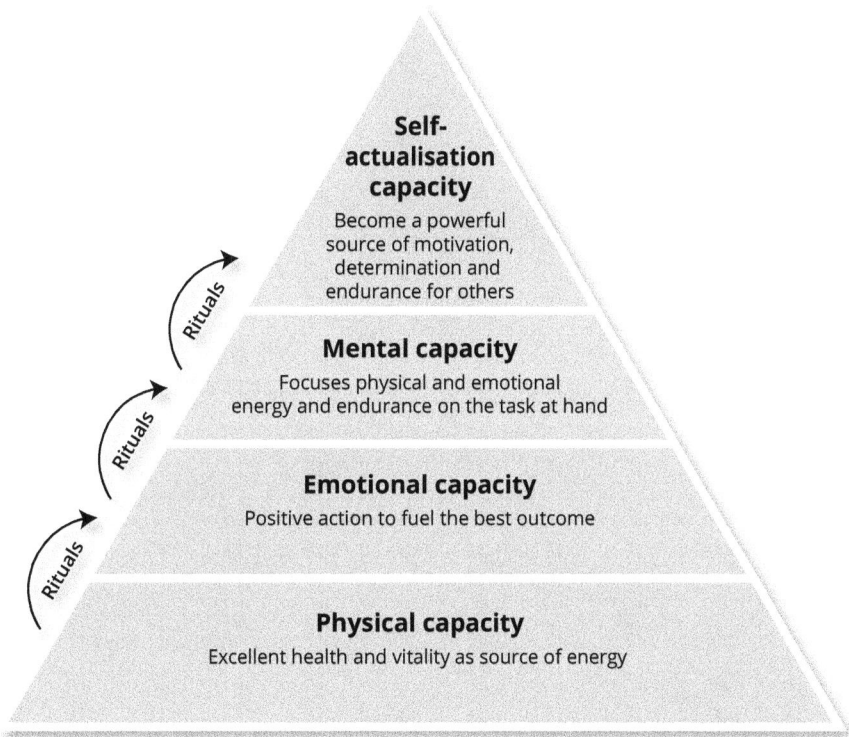

Self-actualisation capacity
Become a powerful source of motivation, determination and endurance for others

Mental capacity
Focuses physical and emotional energy and endurance on the task at hand

Emotional capacity
Positive action to fuel the best outcome

Physical capacity
Excellent health and vitality as source of energy

Rituals

Adapted from: 'The Making of The Corporate Athlete', Jim Loehr, Tony Schwartz, *HBR*, January 2001 .

To achieve and maintain wellbeing in the longer term, and avoid burnout, while undergoing any type of transition (an unavoidable scenario in a VUCA world), leaders would be wise to establish the right routines to help optimise all capacities. This can be done by taking the person as a whole and systematically developing SMART goals, which we'll elaborate on in chapter 11. For now, we can refer to figure 31 to help us explore how to design an action plan to optimise performance, sustainably.

CONSOLIDATE TRANSITION PROCESS

"The curious paradox is that when I accept myself just as I am, then I can change."

Carl Rogers

FACING AND WORKING THROUGH FEARS

We are all wired to welcome anything that is life sustaining and avoid what may be dangerous. The success of rapid and accurate approach–avoidance decisions determines if we live long enough to carry our genes forward to another generation. Because vigilance for danger is a central mechanism of survival, anxiety may be nature's way of ensuring our survival. Indeed, some anxieties appear to be hardwired and linked to our deep past. For instance, fear of spiders, snakes and heights may be linked to the survival needs of our forest-dwelling ancestors. Our complex neural systems have all been sculpted to serve the prime directive of survival, which is triggered in the fight-or-flight circuitry described earlier.

Interestingly, changes to our identity can also be perceived by our neural circuitry as a threat and, hence, as stress. As we saw before, the response to stress results in a range of physiological changes designed to prepare the body for fight or flight. However, as we practise mindfulness and slow down our neurobiological responses – that is, we consciously experience anxiety – we are provided with the opportunity to face and work through our fears. This is particularly significant for people who are about to, or who are undergoing, major career transitions. Some coaches have argued that this is more salient for STEMM professionals as they need time to adjust their perceptions of who they once were and embrace a new reality. However, this applies to any human undergoing major changes to their identity. In the interests of diversity and inclusiveness, it could be argued that it's counterproductive to single out STEMM professionals. The fact is, all humans walk the tightrope of avoiding dramatic changes to their lives and embracing new challenges.

We therefore provide all our coaching clients time and space for reflection on their own sensemaking. We help them see parts of their internal assumptions about the way they have authored their lives and decide what parts they want to change or adjust. Opportunities are given so they can interact with a variety of complex and different viewpoints to what they've been accustomed to. The coach needs to offer the idea that people can author their own path, write their own context, and repeat it as many times as necessary. Formulating new perspectives invariably takes time, and it is therefore crucial that enough time be provided for them to be internalised by the client.

Facilitating the internalisation and consolidation of new perspectives can be aided by a wide variety of coaching interventions, depending on the client's needs and characteristics. As mentioned in a previous chapter, to maintain motivation, these coaching interventions must support the client's *autonomy* (the basic need to be the author of our lives, to have a sense of choice), *relatedness* (the basic need to feel we belong and matter to others) and *competence or mastery* (the basic need to feel effective, to be successful, and to grow). Those three elements are the three nutriments underpinning *self-determination*, which is essential to the psychological wellness of all humans.

NEUROSCIENCE AND STEREOTYPES

Let's consider some stereotypes about the human brain that may be detrimental to any leadership transition process if not properly addressed, especially when working with STEMM leaders.

Female and male brains

The advent of brain imaging technology in the 20th century resulted in the genesis of groundbreaking neuroscience research, as well as unleashing a tidal wave of neuro-nonsense. It seemed that everywhere we turned we'd find a new self-help or management book claiming to have the key to effective leadership, happiness, and even marriage according to neuroscience. The colour-coded brain maps we've all come to love on social media, blogs, and articles are a misrepresentation of real-time brain activity and, sadly, both researchers and the media have played a big part in perpetuating the myth that women are indeed from Venus and men from Mars. Those pesky implicit biases helped push the agenda in brain research to hunt for differences between men and women. Until recently.

The 21st century has brought tremendous advances in brain imaging technology and computational power, and with them, the ability for neuroscientists to interpret data more accurately and to rethink causality links leading to incorrect interpretations. Cognitive neuroscientist and professor Gina Rippon explained in a *New Scientist* article published in March 2019 how recent studies have helped debunk the myth about hardwired differences between the male and female brain. Basically, the evidence suggests that women and men are more similar than different. For sure, biological sex must be considered as one of the variables when studying brains, however there is much more to it: genes, hormones, and the environment (that is, how life experiences shape brain development). Neuroscience alone does not inform our behaviours; it's a multidisciplinary scientific effort.

Since being published in 1992, the pop-psychology book *Men are from Mars, Women are from Venus* has sold over 15 million copies, and has helped popularise a stereotype which had been around long before the author wrote his book: that men and women are distinguished not only by their genitals and sexual characteristics, but also by their brains. Where did this binary perspective come from? This is how the stereotype goes: the hardwired differences in the brains of women and men underpin differences in behaviour, ability, temperament and lifestyle choices. After all, women's brains are slightly smaller than men's; as measured by phrenologists in the 19th century, mapping the skull is indicative of mental faculties and character, right? A stereotype is born.

In addition, the corpus callosum – the bridge of nerve fibres connecting the two halves of the brain – was found to be bigger in women back in the 1980s. This fit the comfortable, binary, pre-existing concept of right and left brains: that the left half of the human brain is responsible for language and analytical/logical types of thinking, whereas the right half handles emotional and creative processes. Women suddenly developed a reputation as excellent multitaskers with greater emotional awareness because of the sup-posedly simultaneous access to both sides of the brain, thanks to the larger corpus callosum. Everyone was convinced that women were indeed from Venus, and men from Mars (interestingly, Venus is the Roman mythology goddess of love and sex).

Not surprisingly, this stereotype fuelled already existing biases about the parts women ought to play in society. If you grow up reading magazines and books, watching television, advertisements and movies, and constantly hearing from your environment that women are more suited to certain tasks (and careers) than men, these become the messages your brain will use to develop your worldview, your mental programming. When not openly addressed, they become implicit biases, and as we've seen to date with gender inequality, the consequences are rather difficult to reverse.

The lopsided brain

According to conventional wisdom, people tend to have a thinking style or a way of doing things that is either right-brained or left-brained. Those who are right-brained are supposed to be intuitive and creative free thinkers. They are big-picture thinkers who experience the world in terms that are descriptive or subjective. Meanwhile, left-brained people tend to be more quantitative and analytical. They pay attention to details and are ruled by logic. This brain dichotomy was popularised in 1979 by the book *Drawing on the Right Side of the Brain*, spawning countless management consulting whitepapers and leadership articles on why STEMM folk viewed the world so differently. In fact, an *MIT Sloan Management Review* article titled "Developing Versatile Leadership" was published in 2003, using the term "lopsided leadership", which continues to be used by many organisational psychologists and coaches.

The truth is, there is no evidence for this, according to studies by neuroscientists at the University of Utah, published in 2013. They looked at the brain scans of more than 1000 people aged 7 to 29 and divided different areas of the brain into 7000 regions to determine whether one side of the brain was more active or connected than the other side. They did uncover patterns showing brain connection that might be strongly left or right lateralised, however they found no evidence that the study participants had a stronger left- or right-sided brain network. Brain scans demonstrated that activity is similar on both sides of the brain, regardless of the person's personality. In fact, if you performed a CT scan, MRI scan, or even an autopsy on the brain of a mathematician and compared it to the brain of an artist, there won't be much of a difference in brain structure.

The authors concluded that the notion of some people being more left-brained or right-brained is more a figure of speech than an anatomically accurate description. This made a great deal of sense to Gina and Rob, who were both good with numbers and could play a musical instrument proficiently. It seems inaccurate then to link these traits to one side of the brain. While scientists don't have all the answers regarding what determines individual personality, it seems unlikely that it's linked to the dominance of one side of the brain. It is time to dispel the lopsided brain stereotype (along with lopsided leadership) and focus on embracing cognitive diversity and the fact that teams are more effective when diverse perspectives are provided and carefully considered.

Rob found the discussion about lopsided brain stereotypes particularly useful as he was able to challenge and discard negative thoughts about his leadership ability. These unhelpful thoughts stemmed from offhand comments made by his CEO about being too much of a "left brain" thinker – that is, analytical. Through cognitive restructuring, and basic psychodynamic education and the potential effects of earlier negative experiences from his youth, Rob was able to identify his unhelpful thought: "I'm not senior leadership material because I approach problems like an engineer, which isn't a desirable business leadership attribute." Equipped with more accurate information about left or right brains, any past or future remarks lost their potency in his mind as he reframed the thought as simply invalid.

For Gina, viewing her brain as neither female nor male, advantaged nor disadvantaged, also helped her restructure and reframe any unhelpful thoughts stemming from this overly popular stereotype. Of course, she would continue to encounter unconscious biases in the workplace for the rest of her career – a fact that's unlikely to change in our lifetime. However, understanding the scientific rationale behind the fallacy bolstered her confidence and control (two of the 4Cs of the mental toughness model in figure 30) strengthening her ability to thrive and experience wellbeing. All this had a significant positive impact in both Rob's and Gina's transition process to more senior leadership roles.

NEW IDENTITIES

Hopefully by now the reader will have developed a different perspective about anyone's ability to learn, grow and evolve. Yes, for some individuals change is more difficult than it is for others. However, with the right approach and environment, it can be achieved, regardless of whether they have a STEMM background or not. As we read earlier, much has been made about the identities of STEMM professionals. Then again, we now know that transitioning to a leadership role, which represents developing a new identity, is likely to be challenging for all individuals, including STEMM professionals.

As such, it's paramount to give each leader the necessary time to internalise new mental programs, and potentially even new values. The length of time will undoubtedly be different for each person. They need time to internalise their new identity and, hence, their self-concept. It should be noted that, as we saw in chapter 6, the expansion of a leader's mental complexity (and hence, maturity) doesn't necessarily involve replacing an old identity with a new one. As leaders grow in mental complexity, they're able to hold more than one identity and manage the paradox. For instance, Gina eventually became very comfortable thinking of herself as a scientist, a corporate executive, a people leader, a wife, a mother, a wine connoisseur and a blogger (she wrote a blog about wines for a local paper).

Figure 32: Self-concept

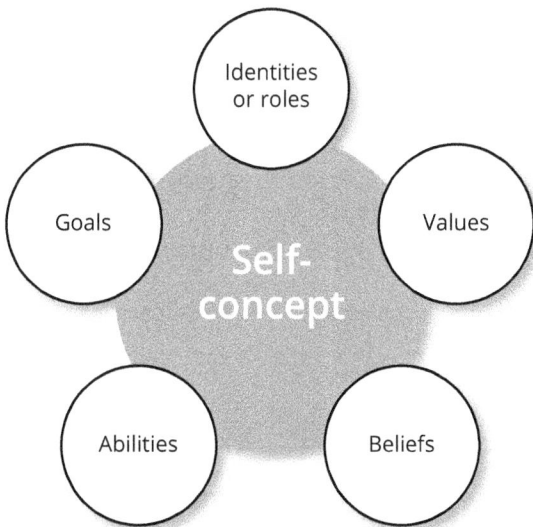

Self-concept is an overarching idea we have about who we are – physically, emotionally, socially, spiritually, and in terms of any other aspects that make up who we are. We form and regulate our self-concept as we grow, based on the knowledge we have about ourselves. It is multidimensional and can be broken down into these individual aspects.[20] The influential self-efficacy researcher Roy Baumeister defined self-concept as the individual's belief about himself or herself, including the person's attributes and who and what the self is.[21] Self-concept is the perspective we have on who we are. Each of us has a unique self-concept, different from the self-concept of others and from their concept of us. This is possibly beyond the scope of this book, however it might be useful to know that when studying the self-concept puzzle, consider several dimensions, which include:

- Self-esteem
- Self-worth
- Self-image (physical)
- Ideal self
- Identities or roles (social)
- Personal traits and qualities.

In coaching, we endeavour to help the client achieve high self-esteem by helping them match their self-concept to their ideal self as closely as possible – see figure 33. Stereotypes impede the achievement of a close match because the ideal self appears to be unattainable.

Figure 33: Matching self-concept to ideal self

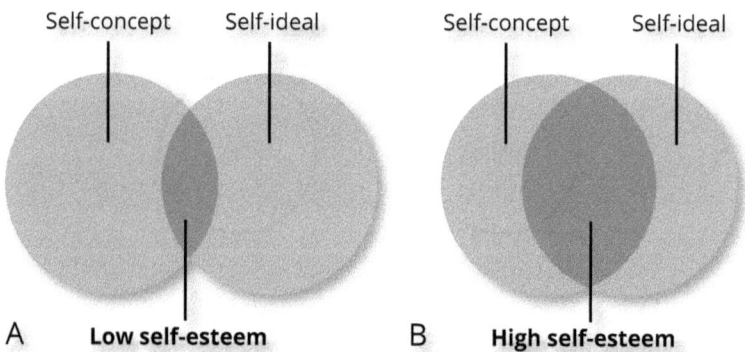

A **Low self-esteem**	B **High self-esteem**

---

20   Marsh, H.W., Xu, K., & Martin, A.J. (2012). "Self-concept: A synergy of theory, method, and application." In K. Harris., S. Graham., & T. Urdan (Eds). *APA Educational Psychology Handbook*. Washington, DC: American Psychological Association.

21   Baumeister, R. (1997). "Identity, Self-Concept, and Self-Esteem". *Handbook Of Personality Psychology*, 681-710. doi: 10.1016/b978-012134645-4/50027-5.

TIME TO STOP (AND BREATHE AND THINK)

We've already spoken about the importance of providing sufficient time for leaders to consolidate the change steps in any the leadership development program. We also spoke at length about the benefits of combining cognitive restructuring tools and mindfulness, also known as MBCT (refer to chapters 8 and 9). The Stop–Breathe–Think, or SBT, tool (shown in figure 34) was designed to bring it all together in one easy place for leaders to refer to daily, or as needed. It can be used as a reflective tool to help consolidate both gradual and step changes along the transition process.

In today's fast-paced workplace, this tool has proven to be very effective and popular with our senior executive clients seeking a "quick intervention" they can apply themselves when faced with a highly stressful situation. It also supports their need for autonomy (especially as they master any of the mindfulness practices).

Step 1 is straightforward, and reminds us to *pause* or *stop* before making a mistake (a wrong decision one can regret later). In step 2 we implement any of the many mindfulness exercises we practise during coaching. These can be as simple as taking 10 deep breaths (what's known as diaphragmatic breathing) or focusing on identifying five sounds, both of which serve to calm the body and the mind. Once we've calmed down we proceed with step 3, which is the cognitive restructuring or reframing, enabling us to find positive solutions to the challenge. The idea is to slow down and think rationally, even when we're faced with problems needing prompt decisions.

Figure 34: Stop–Breathe–Think tool

1. Stop	
What is going on right now?	

2. Breathe	
Take 10 deep breaths	
Notice 5 sounds	

3. Think	
What am I feeling?	
What am I thinking?	
❑Rational ❑Negative ❑Positive	
What do I really want? (No mixed messages)	
What is the best thing to do?	
Advantages	
Disadvantages	
Possible Consequences	
Do I need advice/help?	

CHAPTER 11

THE SCIENCE OF LEADERSHIP

ENABLE SUSTAINABLE CHANGE

"Every success story is a tale of constant adaption, revision and change."

Richard Branson

To be an executive coach is a great privilege. Most often it is much more than helping clients flourish at work. It's also about being their life coach as well. We get to listen to people's life stories, their hopes, their dreams, and their fears. We provide a safe harbour for them during their times of difficulty. We learn about their relationships, who matters, and what matters the most to them. If you show them unconditional positive regard and listen non-judgementally, these amazing people – often highly accomplished individuals with remarkable education and career success – will entrust their future to you.

It's not our job to tell them what to do (although occasionally we might have to ask leading questions). Our job is to shine the light on unforeseen pathways by asking questions. It is to remind them about their unique strengths, to help them come face to face with their fears, and to walk beside them as they take courageous steps to manage and even conquer those fears. Sometimes we are teachers, sometimes cheerers, sometimes scaffolds. At all times, we must be ethical and apply evidence-based methodology. At the heart of evidence-based coaching is goal setting.

The typical "stages of change" models used in coaching are not discussed in this book because both Rob and Gina came to coaching willingly and both were ready for change. It is important for leaders and coaches working with STEMM leaders to be mindful of their readiness for change, as that will inform the initial conversations and subsequent interventions. In addition, more junior leaders might be in the socialised mental complexity phase, still viewing themselves as technical experts. In those cases, the focus is initially on developing foundational people management skills while providing consistent and steadfast support. Depending on the field of STEMM, the organisational context and on the individual's characteristics, this stage could take anywhere between several months and several years. And in all cases, the L&D trainer or the coach, as the case may be, must demonstrate empathy and respect – change is difficult.

It might seem odd to leave goal setting to the last chapter, given its centrality in the coaching process. However, the SCIENCE of Leadership model, as mentioned previously, is not a sequential process. The steps can be applied in any order deemed appropriate by experienced trainers or coaches (except, perhaps, for the first one: self-knowledge development).

SETTING VALUES-BASED GOALS

Goal setting is a powerful process for thinking about your ideal future and for motivating yourself to turn your vision of this future into a reality. Creating goals will help you to enable sustainable change. If you don't already set goals regularly, you may want to start now. As you make this technique part of your life, you'll find your career will become more meaningful, satisfying and successful as it will be part of a holistic and fully integrated, values-driven life plan. Setting goals gives you long-term vision and short-term motivation. Goals serve a directive function; they direct attention towards relevant activities, and away from irrelevant activities. They have an energising function, leading to greater efforts both cognitively and behaviourally. In other words, goals help create focus on what is important as you'll quickly spot the distractions that can easily lead you astray.

By setting sharp, clearly defined goals, you can measure and take pride in the achievement of those goals and you'll see progress in what might have previously looked like a long, pointless grind. You will also raise your self-efficacy (and confidence) as you recognise your own ability and competence in achieving the goals you've set.

You are probably familiar with the concept of setting "S.M.A.R.T." goals. However, you may be surprised to hear there is a "science" behind goal setting which can lead to a richer, more meaningful, and more successful career and life.

The science of goal setting is explained in detail by the leading pioneer in coaching psychology, Professor Anthony Grant, in his breakthrough 2006 book *Evidence Based Coaching Handbook*. Building on the work of other internationally renowned psychologists and executive coaches, he helped "crack the code" to setting and achieving goals successfully. Since then, Grant has continued to add to the extensive body of knowledge on goal setting in coaching practice, as discussed in the 2013 book *Beyond Goals: Effective strategies for coaching and mentoring*.

The first step in setting goals is to consider what you want to achieve in your lifetime, or at least by a distant age in the future. Setting lifetime goals gives you the overall perspective that shapes all other aspects of your life and hence of your decision-making. But, if your goals are too distant, you may also feel overwhelmed by them. You may not know where or how to begin. Therefore, I suggest setting short-, medium- and long-term goals. To give a broad and balanced coverage of all the important areas in your life, it's best to set goals in the following categories:

- **Career** – what level do you want to reach in your career?

- **Financial** – how much do you want to earn, by what stage?

- **Education** – what knowledge do you want to acquire?

- **Family** – how do you want to be seen by members of your family?

- **Artistic** – do you want to achieve artistic goals?

- **Physical/Health** – do you want to achieve health and fitness goals?

- **Pleasure** – how do you want to enjoy yourself?

- **Civic life** – do you want to make the world a better place? How?

Make sure your goals motivate you

When the things that you do and the way you behave match your values, life is usually good – you're satisfied and content. But when these don't align with your personal values, that's when things feel … wrong. This can be a real source of unhappiness.

Core values are the fundamental beliefs of a person (or organisation). These guiding principles guide your behaviour, helping choose between one action and another (even if unconsciously). Ideally, they can help you understand the difference between right and wrong and help you stay on the right path towards fulfilling your goals. However, developing a deep understanding of your core values is not that easy. Depending on the source, you will find that there are more than 50 values, and they include: accountability, accuracy, achievement, adventurousness, altruism, ambition, assertiveness – and that's just the ones starting with "A". Crystallising your thinking further to identify your top 5 or 10 values can be even more powerful. Linking your goals to your top 3 to 5 values will help you stay motivated and on track whenever the going gets tough in reaching your goals.

Figure 35 illustrates how aligning your goals with your values results in motivation. When goals are autonomously chosen and aligned with your values, they are said to be integrated or intrinsically driven, which is the psychological mechanism by which you will be motivated to pursue your goals. On the other hand, if you are forced to carry out goals that are not aligned with your values, they will feel like they are imposed on you; that is, you are *required* to do them. Those goals are said to be external, and you will struggle to be motivated, unless you find a way to link them to your values.

Figure 35: Aligning goals and values

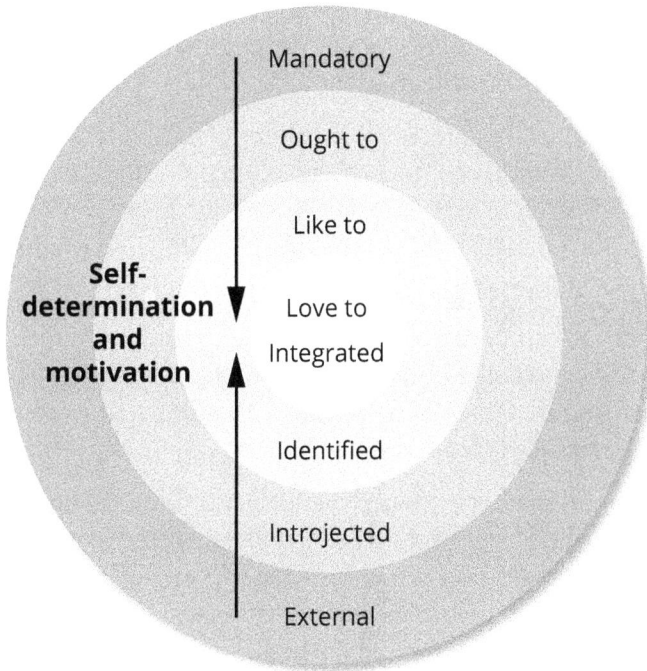

Adapted from: mappalicious.com

Once they have clarified their personal core values, many of my coaching clients request assistance in understanding whether there is a disconnect between their personal and their organisation's values. Sometimes this is not consciously evident to the client; they just know they are lacking motivation or just not delivering, despite their best attempts. Again, skilled coaching can make a significant difference here.

Make your goals tangible

Here are some ideas to help make your goals more achievable:

- **Write your goals down** – this crystallises them and gives them more force. Remember: what doesn't get written down doesn't get done.

- **State each goal as a positive (approach) statement** – express goals positively and constructively, not using avoidance language. "Execute this technique well" is a better goal than "Don't make this silly mistake".

- **Be specific** – set precise goals, putting dates, times, and amounts so you can measure achievement. These are usually referred to as "S.M.A.R.T." goals, and this helps you set priorities, direct attention, and avoid feeling overwhelmed.

- **Develop a goal hierarchy** – break down goals from "higher order" or values-driven (for example, be healthy) targets into "second order" goals and action plans. Keeping goals small and incremental helps direct attention and gives more opportunities for building confidence.

- **Focus on mastery** – when the focus is on mastering a task or learning and developing a solid understanding of a new field, a person feels less stress and feels more positive about achieving the goal. When the focus is on performance (for example, being superior, smarter, or the best at a task), stress levels may become too high and adversely affect goal achievement.

Test your goals for workability

When setting and writing your goals, ask yourself to what extent they:

- Serve a directive function? Do they direct your attention towards relevant activities?

- Energise you? Encourage you to be persistent?

- Lead to the use of task-relevant information and strategies? That is, can they be expressed as S.M.A.R.T. goals?

Goal setting is at the heart of any effective leadership development process because it provides a tangible and real representation of a desired outcome. Without goals, both our careers and our lives would be directionless, following other people's lead.

Setting career goals without a systematic approach can lead to lack of motivation, frustration, unmet career expectations and ultimately an unfulfilling life. It is therefore important for your career goals that you:

1   Set lifetime and holistic goals.

2   Make sure your goals motivate you.

3   Make your goals tangible.

4   Test your goals for workability.

As you make this technique part of your life, you'll find your career will become more meaningful, satisfying and successful as it will be part of a holistic and fully integrated, values-driven life plan.

To help you get started on your journey, whether you work alone on your goals or with a coach, we've included our very popular Goals Matrix in appendix B.

ONGOING CHANGE, LEARNING AND GROWTH: LESSONS FROM GLOBAL CEOs

I felt honoured to be able attend the International CEO Leadership Forum in Boston in October 2018. It was a unique forum for advanced leadership coaches, designed to provide a powerful learning environment for attendees and speakers alike, with world-class executive coaches and C-level leaders from a wide range of industries and sectors globally.

There were several coach-moderated panel discussions with C-level leaders from organisations such as Intel, the World Bank, PayPal, WD-40, the US Federal Government (from previous and current US administrations), Tokidoki, Corent Technology and Harvard Kennedy School. At the end of the two-day conference, we all came together and examined how the world's best executives navigate the overwhelming pressures from conflicting stakeholders and still deliver their best work. These are the 10 key lessons I came away with for enabling sustainable change:

1   Better leadership outcomes can be achieved by making mindfulness practice a central aspect of daily life. Mindfulness helps slow down thinking, providing a *choice point* for balanced decision-making and the development of fundamental leadership capabilities such as emotional intelligence.

2   Top C-level leaders develop nuanced leadership capabilities within long-term coaching engagements (several years). Coaches help leaders by listening empathetically, validating their situation, and by asking the right *what* and *how* questions to elicit deeper insights and hence provide more constructive suggestions when tackling complex issues.

3   Leaders value the safety and confidentiality of the coaching relationship where they can explore their own development needs (for example, improving their emotional intelligence), have a sounding board, and role-play interpersonal scenarios.

4   Top leaders develop coaching skills to help their own teams discover and deliver their best work. Great leaders are great coaches!

5   The greatest leaders are not necessarily charismatic. They are humble, positive, lead by example, have good character, and are resilient.

6   Developing and maintaining high levels of resilience is crucial to leadership effectiveness, which women leaders mentioned more frequently than their male counterparts.

7   Some women leaders spoke about needing to make difficult decisions when their values were at risk of being compromised; for instance, when working within extreme and systemic toxicity. Often, the difficult decision involved moving on at the appropriate time, avoiding cognitive dissonance and thereby preserving their wellbeing.

8   The role played by executive coaches in the decision-making process is seen as invaluable.

9   There are leaders who choose to stay in difficult work environments and influence positive change. They take a calculated risk to pursue a values-based goal linked to their purpose (a goal bigger than them) by first securing the right level of support within the organisation.

10   There is no such thing as "perfect leadership". Top leaders are indeed at the top because they've learnt from previous failures.

In today's bizarre VUCA world, leaders who courageously make difficult, and sometimes unpopular, decisions for the benefit of all stakeholders (not just the shareholders) are nothing short of heroic.

FINAL REMARKS

Despite the Herculean efforts from leadership and coaching scholars and practitioners advancing the application of scientifically based approaches to leadership development, there's still a long way to go before we begin to truly benefit from cognitively diverse, rational leadership.[22] We need to accelerate, significantly, the collaborative efforts and cross-fertilisation between multiple disciplines, such as business schools, coaching psychology, management and leadership, if we are to grow the types of leaders and organisations that will serve the new needs of the 21st century. This call to action will benefit organisations by increasing their ability to innovate, attract and retain talent, become more productive and agile, and achieve greater profitability while being socially responsible.

---

22   Professor Anthony Grant (the world's first coaching psychologist) and his Coaching Psychology Unit at The University of Sydney have been generating research studies for more than 15 years to advance the application of evidence-based methodology to coaching practice. Prior to this, anyone could claim to be a leadership, executive or life coach. Thanks to the efforts of Grant et al., Carol Kaufman and Margaret Moore at the IOC, Harvard Medical School, and many other science-driven coaching pioneers, an increasing number of internationally peer-reviewed studies have been published.

Our case studies are a testament of the power of STEMM leaders in today's irrational world. Gina went on to become the CEO of an Australian biotech company, helping it transition to its commercial phase. Her balanced and values-based decision-making, together with highly sophisticated EI, were instrumental in maximising multiple stakeholders' contributions. These include the board, external commercial partners (multinationals), government agencies and the 150 employees from diverse backgrounds in her organisation. And as a strong advocate of the 17 UN SDGs (figure 1), she's ensuring that the company's strategy and business activities are aligned with the SDGs, not just in principle but in everyday decision-making. She implemented company policies for innovation, gender parity, renewable energy, responsible production targets and a bottom-up initiative to reward employees for achieving and reporting on their individual contribution. Rob is standing out as "a leader of leaders" within the IT industry. As general manager, he's become known for his innovative approach to problem-solving, ability to lead high-performing teams, and for his measured approach in inspiring others to achieve company goals in a sustainable fashion – without taking shortcuts, without being unethical and without burning out. Plus, he's become an active advocate for wellbeing and good health initiatives within the company's Australasian arm.

In a world that seems to glorify the fast-talking, shoot-from-the-hip, charismatic leader, organisations would be wise to address the biases and related systemic obstacles preventing the much-needed new kind of leader from rising to the C-suite and boardroom. To keep neglecting this issue means perpetuating the status quo of deficient cognitive diversity and rational thinking at the leadership table. The benefits of cognitively diverse voices in influential leadership circles, especially STEMM representation, are clear. The fact that this is now a matter of priority reflects the new reality unfolding in the 21st century. This new reality calls for different, evidence-based approaches to developing leaders. The transformation and wellbeing of organisations, communities, nations and the world at large depend on it.

For the highly curious STEMM readers, and for all the intrepid and equally curious CEOs, C-level leaders, and HR leaders, you'll be happy to know you don't have to wait until the next books are published to keep learning more about creating a highly positive and agile organisation. Please visit us at www.proveritas.com.au to find scientifically curated articles and resources on topics to help you improve leadership skills, employee engagement, profitability and growth. Make sure you join our learning community by receiving our quarterly newsletter 'Coaching for Growth' (just find the button at www.proveritas.com.au to sign up). Or if you'd like faster feedback, we'd love to see you at our Facebook, LinkedIn and Twitter pages.

Web:
www.proveritas.com.au

Facebook:
https://www.facebook.com/proveritasgroup/

LinkedIn:
https://www.linkedin.com/company/proveritas-coaching-stemmleaders/

Twitter:
https://twitter.com/ProveritasGroup

ProVeritasGroup

SOME COACHING TOOLS USED IN THE SCIENCE OF LEADERSHIP MODEL

ELEMENT	OBJECTIVES	EXAMPLE TECHNIQUES AND MODELS
Self-knowledge development	Co-create coaching space: safety, confidentiality, positive regard.	Psychometric assessments, e.g. Hogan, VIA Character Strengths.
	Gain self-insight and highlight its value as the basis for life and career decisions (as opposed to external expectations).	Big Five/OCEAN personality test (Costa & McCrae).
		Sociogenic, biogenic, history and organisational context (Khan, Little).
	Understand personality and character strengths to create a road map for continued growth and maximise career outcomes.	Self-awareness 2 × 2 model (Eurich).
		Stereotype content model (Fiske, Cuddy & Glick).
	Learn about blind spots, cognitive biases and stereotypes.	Cognitive restructuring, e.g. cognitive behavioural therapy (CBT); mindfulness.
	Learn to pause, reframe and manage unhelpful thoughts adaptively.	Broaden and Build theory (Fredriksson).
		Growth mindset (Dweck) and Hope Theory (Snyder).
Conceptualise mental complexity	Learn to view life and career transitions as an integral aspect of human development and career progression.	Adaptive complex systems theory and applications to decision-making in leadership.
	Normalise transitions to reduce feelings of isolation.	Questions at relevant phase constructive-developmental transition model (Kegan, Garvey Berger).
	Develop greater mental complexity as core aspect of leadership effectiveness.	Self-determination theory (Ryan & Deci, Spence).
	Celebrate autonomy, promote relatedness, cultivate mastery according to levels of mental complexity.	Create opportunities for self-reflection and seeing the system (Cavanagh, Garvey Berger).
	Foster perspective taking capacity.	Network systems analysis, systems mapping, ripple effect (O'Connor).

ELEMENT	OBJECTIVES	EXAMPLE TECHNIQUES AND MODELS
Investigate leadership styles & capabilities	View leadership as a process whereby leader influences others to achieve goals. Dispel non-scientific stereotypes; right/left brain and male/female brain. Discover most researched leadership models. Maximise leadership strengths to build new leadership capabilities. Develop leadership identity and other-focus.	Discuss most researched leadership models, e.g. transformational, servant and adaptive leadership. Discuss desirable leadership capabilities in the 21st century. Identify capabilities development needs (ProVeritas). New narrative for self-concept and leadership identity. Four-factor leadership model (Cavanagh).
Emotional intelligence enhancement	Learn the importance of emotional intelligence in achieving positive outcomes. Apply innate curiosity, creativity and problem-solving qualities to build EI in STEMM leaders. Develop self-awareness, self-management, social awareness and relationship management dimensions and 12 competencies. Augment sophisticated communication skills for influential leadership.	Mindfulness Course for Executives (ProVeritas). Emotional Intelligence assessments, education and action-learning exercises (Goleman & Boyatzis). Dialogue and crucial communication processes and skills (Patterson). Dynamic authenticity model to normalise negative emotions and scaffold the creation of a new identity (Jay & Grant).

ELEMENT	OBJECTIVES	EXAMPLE TECHNIQUES AND MODELS
Nurture resilience & wellbeing	Build effective and positive coping strategies to deal with challenges of transition. Enable application of variety of coping strategies in a flexible manner depending on the specific challenge, and then use corrective feedback to adjust those strategies. Develop psychological flexibility.	Physical, emotional and mental capacity training, e.g. MTQ48. Psychodynamics, positive psychology interventions: gratitude, purpose, meaning. Psychological flexibility, mindfulness practice and Acceptance & Commitment Therapy (ACT) at work (Harris). Tools to create new habits. The making of a corporate athlete (Loehr & Schwartz).
Consolidate transition process	Promote positive evaluations of the past, present and future for clearer self-concept, identity and coherence. Allow enough time and space to work through the process of transition. Cultivate motivation and wellbeing through autonomy, mastery and relatedness.	Integrating multiple identities/selves. Attachment theory and relationships (Cozolino). Review generational impact on systems (Bowen). Neuroscience research and coaching. Contrast, compare and integrate with coachee's own expectations. Self-determination theory (Ryan, Deci, Spence).

189

ELEMENT	OBJECTIVES	EXAMPLE TECHNIQUES AND MODELS
Enable sustainable change	Achieve sustainable change via values-based, fully integrated goals, strategies and solutions. Understand balance between self-compassion and resilience. Accept that change is messy while staying on track. Replenish internal and external resources.	Stages of change (Prochaska). Solution-focused, values-based *what* and *how* questions (Grant). Balanced goals matrix with all major areas of life (not just career). SMART goals and action plans (Grant). House of change (Grant). The 7Rs for lasting change (Harris). Technology-based support of lifelong learning and mentoring (ProVeritas).

APPENDIX B

GOALS MATRIX

Work	Relationships
My Values:	My Values:
Short-term goals:	Short-term goals:
Medium-term goals:	Medium-term goals:
Long-term goals:	Long-term goals:

Play	Health
My Values:	My Values:
Short-term goals:	Short-term goals:
Medium-term goals:	Medium-term goals:
Long-term goals:	Long-term goals:

ProVeritas Group
Coaching for Growth
www.proveritas.com.au

193

ACKNOWLEDGEMENTS

Gratitude is a wonderful value to cultivate. Expressing gratitude is one of the first positive coaching exercises we do with our leaders – it generates positivity and helps us focus on who and what is important in life. Practising gratitude is also a great way of cultivating mindfulness. For these reasons, and many more, I'm thrilled to arrive at this stage of the long journey that is to write a book like this one. And at the risk of sounding like an actor giving her speech after receiving an Oscar, I'm grateful to my tribe – those who supported, encouraged and coached me throughout this arduous process. I also get to thank my extended community of family, friends and colleagues from all over the world.

No words can express my profound gratitude to my best friend, life partner and husband, Darrell C. Campbell, for creating "a calm space" in our lives for me to spend countless hours over the past three years doing research, conducting interviews, travelling, analysing data and typing. Not once did you complain about being left alone with our son during those hours while I was away or locked up in my home office – thinking and typing. Quite often

I'd be annoyed at my computer due to never-ending connectivity glitches that seem to plague all residents of the Upper North Shore of Sydney – and yet you'd always find a way to get me back online as quickly as possible. You always encouraged me to stay strong when "the going got tough". There's no question about it, career women need to choose their partners wisely if they're to succeed in any given endeavour. And I chose mine wisely.

To my gorgeous son Isaak J. Campbell, my loyal cheerleader, thank you for being supportive, positive and wise beyond your adolescent years. I appreciated every word of encouragement you gave me – being told that I'm the best mum in the world is the most significant and rewarding feedback I've ever received. You weren't praising me to ask for something! During these past few years you've shown great courage as you did all your schoolwork, house chores and sports activities with determination, despite my occasional moments of forgetfulness. I know it wasn't always easy, but you were determined to do your part, as a member of our little clan, so that I wouldn't be distracted from the task at hand. And I loved it that you'd often do your homework, or simply read a book, on the couch in my home office, just so we could be near each other. Thank you, darling son!

I'm forever grateful to my parents for teaching me from a young age to love learning and to help others. To my amazing mum Aida, your inner strength, determination and discipline are nothing short of inspiring – you're a role model for all women. Thank you for showing me the importance of standing up for my myself, and to be authentic. I wish my late dad Victor were here to celebrate this moment. I'm glad I had the chance to tell him how grateful I am for teaching me to love knowledge. I'll never forget his words when growing up: "People might take away your material posses-sions. But nobody can take away your knowledge. That's all yours." Thank you dad.

To my friends and family near and far, thank you for checking in on me whenever I went "radio silent", and even when I didn't. Your love and support mean the world to me. To my dearest friend Ivonne Jimenez, I'm very grateful for your enduring and wonderful sense of humour, positivity and encouragement. I hope you know how much I value and love you. Our friendship is proof that introverts and extroverts can be close friends – it's a matter of looking for common ground, being accepting, kind (especially with time and positivity) and forgiving.

I'm deeply grateful to my schoolteachers, university professors, coaches, mentors, colleagues, and previous managers – they've all taught me something valuable. Your impact in my life is reflected in the coaching, education and mentoring work I do. There are too many names to list here and I don't want to run the risk of leaving anyone out.

Thank you also to all the wonderful STEMM professionals and leaders who gave me their time to be interviewed. I sincerely appreciate your willingness and courage to speak openly about your successes as well as your trials and tribulations. And there is no question about it, it's significantly more difficult to be a woman in STEMM than it is for men. I purposely left the intricacies of the topic out of this book – it deserves its own space, full attention and separate book.

Thank you to my remarkable coaching clients, for having the courage to want to flourish and not settle for "just existing". It is an honour and a privilege to accompany you on your journey to greater success in life, whichever way you choose to define it. Each one of you has enriched my life and I'm humbled to have been part of yours.

I also wish to thank my editor and publisher Michael Hanrahan and his wonderful team. Thank you for helping me untangle my ideas so they would make sense to the reader. You were also another positive and calming force, especially throughout the latter stages. I sincerely appreciate your ability to keep me on track and "pull me in" whenever I went off on a conceptual tangent.

Last but not least, I'd like to thank the leadership at CustomLinc Pty Ltd for sponsoring the production of this book. Choosing to self-publish my first book wouldn't have been possible without your generous assistance. Your company is a shining example of what can be achieved when the SCIENCE of leadership is applied throughout the business. I'm confident you'll continue to go from strength to strength, demonstrating that STEMM leaders can transform the lives of employees, their families, their communities, and the world at large.

BIBLIOGRAPHY

314 Action. (2019). Retrieved 1 July 2019, from https://www.314action.org/

ABC Q&A. (2019). Q&A Science Special. Retrieved 26 July 2019, from https://www.abc.net.au/qanda/2019-17-06/11191192

Achakulwisut, P. (2017). Why Are Scientists So Averse to Public Engagement? [Blog]. Retrieved from https://blogs.scientificamerican.com/guest-blog/why-are-scientists-so-averse-to-public-engagement/

Al-Saleh, Y. Why Engineers Make Great CEOs. INSEAD. Retrieved from https://knowledge.insead.edu/blog/insead-blog/why-engineers-make-great-ceos-3318

Arden, J. (2013). [Blog]. *Rewire Your Brain: Think your way to a better life*. Hoboken, N.J.: John Wiley & Sons Inc.

Atkins, P., & Parker, S. (2012). Understanding Individual Compassion in Organizations: The Role of Appraisals and Psychological Flexibility. *Academy of Management Review*, 37(4), 524-546. doi: 10.5465/amr.2010.0490

Australian Government Department of Industry, Innovation and Science. (2017). *Australia's National Science Statement*. ACT: Australian Government. Retrieved from https://publications.industry.gov.au/publications/nationalsciencestatement/index.html

Australian Government Office of the Chief Scientist. (2016). *Australia's STEM Workforce: Science, Technology, Engineering and Mathematics* (pp. 2-158). Canberra, Australia: Commonwealth of Australia.

Baard, P. (2019). Intrinsic Need Satisfaction in Organizations: A Motivational Basis of Success in For-Profit and Not-for-Profit Settings. In E. Deci & R. Ryan, *Handbook of Self-Determination Research* (pp. 255-275). Rochester, NY: The University of Rochester Press.

Barends, E. (2015). *In Search of Evidence: Empirical findings and professional perspectives on evidence-based management* (PhD). Amsterdam: VU University.

Bartels, M., van Weegen, F., van Beijsterveldt, C., Carlier, M., Polderman, T., Hoekstra, R., & Boomsma, D. (2012). The Five Factor Model of Personality and Intelligence: A twin study on the relationship between the two constructs. *Personality and Individual Differences*, 53(4), 368-373. doi: 10.1016/j.paid.2012.02.007

Bassuk, A. (2017). Leadership Crisis/Crisis Leadership. Retrieved from https://www.huffpost.com/entry/leadership-crisiscrisis-leadership_b_59dfd639e4b09e31db97579e

Baumeister, R. (1997). Identity, Self-Concept, and Self-Esteem. Handbook of Personality Psychology, 681-710. doi: 10.1016/b978-012134645-4/50027-5 Benefits of Leadership Coaching. (2019). Retrieved 7 July 2019, from http://www.proveritas.com.au/leadership-coaching

Berger, J. (2012). *Changing on the Job: Developing Leaders for a Complex World* (pp. 49-177). Stanford, CA: Stanford Business Books.

Berger, J., & Johnston, K. (2015). *Simple Habits for Complex Times: Powerful Practices for Leaders* (pp. 88-234). Stanford, CA: Stanford Business Books.

Bertolero, M., & Bassett, D. S. (2019). How the Mind Emerges from the Brain's Complex Networks. Scientific American. Retrieved from https://www. scientificamerican.com/article/how-the-mind-emerges-from-the-brains-complex-networks/

Birshan, M., Meekin, T., & Strovink, K. (2017). What Makes a CEO Exceptional? *Mckinsey Quarterly Five Fifty*.

Bjerkedal, T., Kristensen, P., Skjeret, G., & Brevik, J. (2007). Intelligence Test Scores and Birth Order Among Young Norwegian Men (conscripts) Analyzed Within and Between Families. *Intelligence, 35*(5), 503-514. doi: 10.1016/j.intell.2007.01.004

Bogenschneider, B. (2016). Leadership Epistemology. *Creighton Journal of Interdisciplinary Leadership*, 2(2), 24. doi: 10.17062/cjil.v2i2.37

Boniwell, I., Kauffman, C., & Silberman, J. (2014). The Positive Psychology Approach to Coaching. In E. Cox, T. Bachkirova & D. Clutterbuck, *The Complete Handbook of Coaching* (2nd ed., pp. 157-169). London: SAGE Publications.

Bourne, A., & Whybrow, A. (2019). Using Psychometrics in Coaching. In S. Palmer & A. Whybrow, *Handbook of Coaching Psychology: A Guide for Practitioners* (2nd ed., pp. 512-526). New York: Routledge.

Boyatzis, R., Goleman, D., & Rhee, K. (2000). Clustering Competence in Emotional Intelligence: Insights from the emotional competence inventory. In R. Bar-On & J. Parker, *Handbook of Emotional Intelligence* (pp. 343-362). San Francisco: Jossey-Bass.

Boyle, G. (1995). Myers Briggs Type Indicator (MBTI): Some Psychometric Limitations. *Australian Psychologist, 30*(1), 71-74. Retrieved from https://doi.org/10.1111/j.1742-9544.1995.tb01750.x

Boys, K. (2018). *The Blind Spot Effect: how to stop missing what's right in front of you.* Boulder, CO: Sounds True Inc.

Bremer, M. (2018). *Developing a Positive Culture Where People and Performance Thrive.* New York: Motivational Press, Incorporated.

Cain, S. (2013). *Quiet: The Power of Introverts in a World That Won't Stop Talking.* London: Penguin Books.

Cain, S. Quiet Revolution: Unlocking the Power of Introverts. Retrieved 30 July 2019, from https://www.quietrev.com/

Canaday, S. (2017). Cognitive Diversity. [Blog]. Retrieved from https://www. psychologytoday.com/au/blog/you-according-them/201706/cognitive-diversity

Cavanagh, M. (2013). The Coaching Engagement in the Twenty-first Century: New Paradigms for Complex Times. In S. David, D. Clutterbuck & D. Megginson, *Beyond Goals: Effective Strategies for Coaching and Mentoring* (pp. 151-183). Surrey, England: Gower Publishing Ltd.

Chamorro-Premuzic, T. (2013). Why Do So Many Incompetent Men Become Leaders?. *Harvard Business Review*. Retrieved from https://hbr.org/2013/08/why-do-so-many-incompetent-men

Charlton, E. (2019). World Economic Forum. [Blog]. Retrieved from https://www.weforum.org/agenda/2019/05/new-zealand-is-publishing-its-first-well-being-budget/

Chinsky Matuson, R. (2017). You're Kidding, Right? 50% of New Hires Fail. [Blog]. Retrieved from https://www.linkedin.com/pulse/youre-kidding-right-50-new-hires-fail-roberta-chinsky-matuson/

Christakis, N., & Fowler, J. (2012). Social Contagion Theory: Examining dynamic social networks and human behavior. *Statistics In Medicine*, *32*(4), 556-577. doi: 10.1002/sim.5408

Collier, P. (2019). *The Future of Capitalism: Facing the New Anxieties*. [S.l.]: Penguin Books.

Cook, J., Oreskes, N., Doran, P., Anderegg, W., Verheggen, B., & Maibach, E. et al. (2016). Consensus on consensus: a synthesis of consensus estimates on human-caused global warming. *Environmental Research Letters*, *11*(4), 048002. doi: 10.1088/1748-9326/11/4/048002

Coutu, D., & Kauffman, C. (2009). What Can Coaches Do For You?. *Harvard Business Review*. Retrieved from https://hbr.org/2009/01/what-can-coaches-do-for-you

Covey, S. (1989). *The Seven Habits of Highly Effective People* (1st ed.). New York: Simon & Schuster.

Covey, S. (2004). *The 8th Habit: from effectiveness to greatness*. London: Simon & Schuster.

Cozolino, L. (2014). *The Neuroscience of Human Relationships: Attachment and the Developing Social Brain* (2nd ed.). New York, NY: W.W. Norton & Company Inc.

Cozolino, L. (2017). *The Neuroscience of Psychotherapy: Healing the Social Brain* (3rd ed.). New York: Norton & Company Inc.

Crane, B., & Hartwell, C. (2018). Developing Employees' Mental Complexity: Transformational Leadership as a Catalyst in Employee Development. *Human Resource Development Review*, *17*(3), 234-257. doi: 10.1177/1534484318781439

Cribb, J., & Sari, T. (2010). *Open Science* (pp. 1-14, 202-211). Collingwood: CSIRO.

CSIRO and Bureau of Meteorology, Australian Government. (2018). *State of the Climate 2018*. Retrieved from http://www.bom.gov.au/state-of-the-climate/

CSIRO and NAB. (2019). *Australian National Outlook 2019*. Retrieved from https://www.csiro.au/en/Showcase/ANO

Dabke, D. (2016). Impact of Leader's Emotional Intelligence and Transformational Behavior on Perceived Leadership Effectiveness: A Multiple Source View. *Business Perspectives And Research*, *4*(1), 27-40. doi: 10.1177/2278533715605433

De Meuse, K., Dai, G., & Lee, R. (2009). Evaluating the effectiveness of executive coaching: beyond ROI?. *Coaching: An International Journal of Theory, Research and Practice*, *2*(2), 117-134. doi: 10.1080/17521880902882413

Deloitte Touche Tohmatsu Limited. (2016). *Global Human Capital Trends: The new organisation, different by design*. UK: Deloitte University Press. Retrieved from https://www2.deloitte.com/content/dam/Deloitte/global/Documents/HumanCapital/gx-dup-global-human-capital-trends-2016.pdf

Deloitte Touche Tohmatsu Limited. (2019). *Deloitte Global Human Capital Trends: Leading the social enterprise, reinvent with a human focus*. UK: Deloitte Insights. Retrieved from https://www2.deloitte.com/content/dam/insights/us/articles/5136_HC-Trends-2019/DI_HC-Trends-2019.pdf

Dettmer, O. (2011). Minds like machines – Technocrats. *The Economist*, (International Print Edition). Retrieved from https://www.economist.com/international/2011/11/19/minds-like-machines

DeWall, C., & Bushman, B. (2011). Social Acceptance and Rejection. *Current Directions In Psychological Science*, 20(4), 256-260. doi: 10.1177/0963721411417545

Dweck, C. (2017). *Mindset – Updated Edition: Changing The Way You think To Fulfil Your Potential* (6th ed.). London: Robinson.

Edelman Holdings Inc. (2018). *2018 Edelman Trust Barometer Annual Global Report*. Edelman Holdings. Retrieved from https://www.edelman.com/research/2018-edelman-trust-barometer

Edelman Holdings Inc. (2019). *2019 Edelman Trust Barometer Annual Global Report*. Edelman Holdings. Retrieved from https://www.edelman.com/trust-barometer

Egan, D. (2018). Here is what it takes to become a CEO, according to 12,000 LinkedIn profiles. [Blog]. Retrieved from https://business.linkedin.com/talent-solutions/blog/trends-and-research/2018/what-12000-ceos-have-in-common

Emre, M. (2019). *The Personality Brokers: The Strange History of Myers-Briggs and the Birth of Personality Testing*. [S.l.]: Anchor.

Encyclopaedia Brittanica. (2019). Bill Gates: American Computer Programmer, Businessman, and Philanthropist. (2019). [Blog]. Retrieved from https://www.britannica.com/biography/Bill-Gates

Eurich, T. (2017). *Insight: The Power of Self-Awareness in a Self-Deluded World*. London: CPI Group.

Eurich, T. (2018). What Self-Awareness Really Is (and How to Cultivate It). [Blog]. Retrieved from https://hbr.org/2018/01/what-self-awareness-really-is-and-how-to-cultivate-it?autocomplete=true

Eva, N., Robin, M., Sendjaya, S., van Dierendonck, D., & Liden, R. (2019). Servant Leadership: A systematic review and call for future research. *The Leadership Quarterly*, 30(1), 111-132. doi: 10.1016/j.leaqua.2018.07.004

Fallacy Man (pseudonym). (2016). The Hierarchy of Evidence: Is the study's design robust?. (2016). [Blog]. Retrieved from https://thelogicofscience.com/2016/01/12/the-hierarchy-of-evidence-is-the-studys-design-robust/

Fernandez-Araoz, C., Roscoe, A., & Aramaki, K. (2017). Turning Potential to Success: The Missing Link in Leadership Development (pp. 86-93). *Harvard Business Review*, Nov-Dec.

Fiske, S., Cuddy, A., Glick, P., & Xu, J. (2002). A Model of (Often Mixed) Stereotype Content: Competence and warmth respectively follow from perceived status and competition. *Journal Of Personality And Social Psychology*, *82*(6), 878-902. doi: 10.1037//0022-3514.82.6.878

Fonseca, C. (2013). *Quiet Kids: Help Your Introverted Child Succeed in an Extroverted World*. Sourcebooks.

Foulk, T., Woolum, A., & Erez, A. (2016). Catching Rudeness is Like Catching a Cold: The contagion effects of low-intensity negative behaviors. *Journal of Applied Psychology*, *101*(1), 50-67. doi: 10.1037/apl0000037

Frangos, C. (2018). 3 Transitions Even the Best Leaders Struggle With. *Harvard Business Review*. Retrieved from https://hbr.org/2018/07/3-transitions-even-the-best-leaders-struggle-with

Funk, C., & Kennedy, B. (2019). *Public Confidence in Scientists has Remained Stable for Decades*. Pew Research Center. Retrieved from https://www.pewresearch.org/fact-tank/2019/03/22/public-confidence-in-scientists-has-remained-stable-for-decades/

Garvey Berger, J. (2006). Adult Development Theory and Executive Coaching Practice. In D. Strober & A. Grant, *Evidence Based Coaching Handbook* (pp. 77-102). Hoboken, New Jersey: John Wiley & Sons Inc.

Gast, A. (2015). Why Business Leaders Should Think Like Scientists. [Blog]. Retrieved from https://www.weforum.org/agenda/2015/01/why-business-leaders-should-think-like-scientists/

Gates, B. (2008). *Business @ the speed of thought*. Harlow: Penguin.

Gibson, K. (2016). Rethinking the Leadership Industry. [Blog]. Retrieved from https://www.hks.harvard.edu/research-insights/policy-topics/public-leadership-management/rethinking-leadership-industry

Goldenberg, A., Garcia, D., Halperin, E., Zaki, J., Kong, D., Golarai, G., & Gross, J. (2019). Beyond Emotional Similarity: The role of situation-specific motives. *Journal Of Experimental Psychology: General*. doi: 10.1037/xge0000625

Goleman, D. (2013). Focus: The Hidden Driver of Excellence (1st ed., pp. 98-115). London, UK: Bllomsbury.

Goleman. D. (1995). *Working with Emotional Intelligence*. London: Bloomsburg.

Goleman, D. (2006). *Emotional Intelligence*. New York: Bantam Books.

Goleman, D. (2014). *Focus*. Harper Collins USA.

Goleman, D., & Boyatzis, R. (2017). Emotional Intelligence Has 12 Elements. Which Do You Need to Work On?. *Harvard Business Review*. Retrieved from https://hbr.org/2017/02/emotional-intelligence-has-12-elements-which-do-you-need-to-work-on

Goleman, D., Boyatzis, R., & McKee, A. (2002). *Primal leadership*. Boston, Mass: Harvard Business School Press.

Graf, N., Fry, R., & Funk, C. (2018). *7 Facts About the STEM Workforce*. Pew Research Center.

Grant, A. (2006). An Integrative Goal-Focused Approach to Executive Coaching. In D. Stober & A. Grant, *Evidence Based Coaching Handbook*. Hokoben, New Jersey: John Wiley & Sons Inc.

Grant, A. (2012). ROI is a Poor Measure Of Coaching Success: Towards a more holistic approach using a well-being and engagement framework. *Coaching: An International Journal Of Theory, Research And Practice*, 5(2), 74-85. doi: 10.1080/17521882.2012.672438

Grant, A. (2016). *Originals: How Non-conformists Change the World*. London: Penguin Random House.

Grant, A. (2016). The Third 'Generation' of Workplace Coaching: creating a culture of quality conversations. *Coaching: An International Journal of Theory, Research and Practice*, 10(1), 37-53. doi: 10.1080/17521882.2016.1266005

Grant, A., & Greene, J. (2004). *It's Your Life. What Are You Going To Do About It?* (2nd ed.). UK: Momentum.

Hanson, R. (2018). *Resilient*. Random House.

Harney, M., John, R., & Valdivieso, M. (2019). *Meet the Missing Ingredient in Successful Sales Transformations: Science*. McKinsey & Company. Retrieved from https://www.mckinsey.com/business-functions/marketing-and-sales/our-insights/meet-the-missing-ingredient-in-successful-sales-transformations-science

Harris, R. (2009). *ACT Made Simple*. Oakland, CA: New Harbinger Publications Inc.

Harris, R. (2010). *The Confidence Gap: from fear to freedom*. Haymarket, NSW: Penguin.

Heifetz, R., Grashow, A., & Linsky, M. (2009). *The Practice of Adaptive Leadership*. Boston, Mass.: Harvard Business Press.

Helliwell, J., Layard, R., & Sachs, J. (2019). *World Happiness Report 2019*. New York: Sustainable Development Solutions Network. Retrieved from https://worldhappiness.report/ed/2019/

Hendriksen, E. *How To Be Yourself: Quiet Your Inner Critic and Rise Above Social Anxiety*. New York, NY: St Martin's Press.

Hodgkinson, G. (2014). The Politics of Evidence-Based Decision Making. In D. Rousseau, *The Oxford Handbook of Evidence-Based Management* (pp. 404-419). Oxford: Oxford University Press.

Hougaard, R. (2018). The Real Crisis in Leadership. [Blog]. Retrieved from https://www.forbes.com/sites/rasmushougaard/2018/09/09/the-real-crisis-in-leadership/#4ce09b423ee4

Howard, C. (2005). Blueprint for a Great Leader. [Blog]. Retrieved from http://www.pharmexec.com/blueprint-great-leader

Hurd, J. (2009). Development Coaching: Helping scientific and technical professionals make the leap into leadership. *Global Business and Organizational Excellence*, 28(5), 39-51. doi: 10.1002/joe.20277

Hurd, J. (2013). Development Coaching: Helping scientific and technical professionals make the leap into leadership. *IEEE Engineering Management Review, 41*(1), 53-64. doi: 10.1109/emr.2013.6489839

Hurd, J., & Juri, T. (2005). Coaching the Scientific and Technical Professional. *The International Journal Of Coaching In Organizations, 3*(1), 31-38.

Ignatius, A. (2014). The Best-Performing CEOs in the World. *Harvard Business Review*. Retrieved from https://hbr.org/2014/11/the-best-performing-ceos-in-the-world

Instaread. (2016). *Summary, Analysis & Review of Christopher H. Achen's & Larry M. Bartels's Democracy for Realists*. San Francisco: Instaread.

IPCC. Intergovernmental Panel on Climate Change. Retrieved 10 July 2019, from https://www.ipcc.ch/

Jarrett, C. (2017). 5 Reasons It's So Hard to Think Like a Scientist. *BPS Research Digest*. Retrieved from https://digest.bps.org.uk/2017/06/20/5-reasons-its-so-hard-to-think-like-a-scientist/

Jay, J., & Grant, G. (2017). *Breaking Through Gridlock: The Power of Conversation in a Polarized World*. Oakland, CA: Berret-Koehler Publishers, Inc.

Kahn, M. (2014). *Coaching On The Axis: Working with Complexity in Business and Executive Coaching*. London: Karnac Books Ltd.

Kahneman, D., & Egan, P. (2011). *Thinking, fast and slow*. New York: Random House Audio.

Kaiser, R. (2003). Developing versatile leadership. *MIT Sloan Management Review*, (Vol 44, No 4), 19-26. Retrieved from https://sloanreview.mit.edu/article/developing-versatile-leadership/

Kaplan, S., & Sorensen, M. (2017). Are CEOs Different? Characteristics of Top Managers. *SSRN Electronic Journal*. doi: 10.2139/ssrn.2747691

Kaplan, S., Klebanov, M., & Sorensen, M. (2012). Which CEO Characteristics and Abilities Matter?. *The Journal Of Finance, 67*(3), 973-1007. doi: 10.1111/j.1540-6261.2012.01739.x

Karp, P. (2018). A recent history of Australia's banking scandals. [Blog]. Retrieved from https://www.theguardian.com/australia-news/ng-interactive/2018/apr/19/a-recent-history-of-australias-banking-scandals

Kashdan, T., DeWall, C., Masten, C., Pond, R., Powell, C., & Combs, D. et al. (2014). Who Is Most Vulnerable to Social Rejection? The Toxic Combination of Low Self-Esteem and Lack of Negative Emotion Differentiation on Neural Responses to Rejection. *Plos ONE, 9*(3), e90651. doi: 10.1371/journal.pone.0090651

Kaufman, C. (2006). Positive Psychology: The Science at the Heart of Coaching. In D. Stober, *Evidence Based Coaching Handbook*. Hokoben, New Jersey: John Wiley & Sons Inc.

Kegan, R. (1994). In Over Our Heads: The Mental Demands of Modern Life (1st ed.). Cambridge, Massachusetts: Harvard University Press.

Kegan, R., & Lahey, L. (2009). *Immunity to Change*. Boston, Mass.: Harvard Business Press.

Kegan, R. (1994). *In Over Our Heads: The mental demands of modern life*. Harvard University Press.

Kelloway, E., Turner, N., Barling, J., & Loughlin, C. (2012). Transformational Leadership and Employee Psychological Well-being: The mediating role of employee trust in leadership. *Work & Stress, 26*(1), 39-55. doi: 10.1080/02678373.2012.660774

Kennedy, B., Hefferon, M., & Funk, C. (2018). *Half of Americans Think Young People Don't Pursue STEM Because It Is Too Hard*. Pew Research Center. Retrieved from https://www.pewresearch.org/fact-tank/2018/01/17/half-of-americans-think-young-people-dont-pursue-stem-because-it-is-too-hard/

Kets De Vries, M. (2014). *Mindful Leadership Coaching: Journeys Into the Interior* (pp. 4-16, 185-200). New York, NY: Palgrave Macmillan.

Kobayashi-Solomon, E. (2019). A Historic Inflection Point In Capitalism's Battle Against Climate Change. (2019). [Blog]. Retrieved from https://www.forbes.com/sites/erikkobayashisolomon/2019/04/26/historic-inflection-point-mankinds-battle-against-climate-change/#5c2394757829

Knowledge@Wharton. (2018). Personality Puzzler: Is There Any Science Behind Myers-Briggs?. (2018). [Blog]. Retrieved from https://knowledge.wharton.upenn.edu/article/does-the-myers-briggs-test-really-work/

Kotcher, J., Myers, T., Vraga, E., Stenhouse, N., & Maibach, E. (2017). Does Engagement in Advocacy Hurt the Credibility of Scientists? Results from a Randomized National Survey Experiment. *Environmental Communication, 11*(3), 415-429. doi: 10.1080/17524032.2016.1275736

Lamm, A., Shoulders, C., Roberts, G., Irani, T., Snyder, L., & Brendemuhl, J. (2012). The Influence of Cognitive Diversity on Group Problem Solving Strategy. *Journal Of Agricultural Education, 53*(1), 18-30. doi: 10.5032/jae.2012.01018

Lane, D., Kahn, M., & Chapman, L. (2019). Adult Learning as an Approach to coaching. In S. Palmer & A. Whybrow, *Handbook of Coaching Psychology: A Guide for Practitioners* (2nd ed., pp. 369-380). New York, NY: Routledge.

Laney, M. (2002). *The Introvert Advantage: How Quiet People Can Thrive in an Extrovert World*. New York: Workman Publishing.

Lawrence, P., & Whyte, A. (2013). Return on Investment In Executive Coaching: A practical model for measuring ROI in organisations. *Coaching: An International Journal Of Theory, Research And Practice, 7*(1), 4-17. doi: 10.1080/17521882.2013.811694

Little, B. (2011). Personal Projects Analysis. Retrieved 25 July 2019, from http://www.brianrlittle.com/Topics/research/personal-projects-analysis/?doing_wp_cron=1564023169.6640150547027587890625

Little, B. (2014). Personal Projects and Free Traits: On integration in personality science. *Personality And Individual Differences, 60*, S17. doi: 10.1016/j.paid.2013.07.379

Little, B. (2016). *Me, Myself, and Us: The Science of Personality and the Art of Wellbeing*. New York, N.Y.: Public Affairs.

Little, B., & Coulombe, S. (2015). Personal Projects. *International Encyclopedia of the Social & Behavioural Sciences, 17*, 757-765. Retrieved from http://www.ippanetwork. org/wp-content/uploads/2015/09/Personal-projects.pdf

Lounsbury, J., Foster, N., Patel, H., Carmody, P., Gibson, L., & Stairs, D. (2011). An investigation of the personality traits of scientists versus nonscientists and their relationship with career satisfaction. *R&D Management, 42*(1), 47-59. doi: 10.1111/j.1467-9310.2011.00665.x

Lytkina Botelho, E., Rosenkoetter Powell, K., & Wong, N. (2018). The Fastest Path to the CEO Job, According to a 10-Year Study. *Harvard Business Review*.

Lytkina Botello, E., Rosenkoetter Powell, K., Kincaid, S., & Wang, D. (2017). What Sets Successful CEOs Apart. *Harvard Business Review*.

Marsh, H.W., Xu, K., & Martin, A.J. (2012). Self-concept: A synergy of theory, method, and application. In K. Harris., S. Graham., & T. Urdan (Eds). *APA Educational Psychology Handbook*. Washington, DC: American Psychological Association.

McCann, J. (2016). Traits and Trends of Australia's Prime Ministers, 1901 to 2015: a quick guide. [Blog]. Retrieved from https://www.aph.gov.au/About_Parliament/ Parliamentary_Departments/Parliamentary_Library/pubs/rp/rp1516/Quick_Guides/ AustPM

McKinsey & Company. (2017). What's Missing in Leadership Development?.. Retrieved from https://www.mckinsey.com/featured-insights/leadership/whats-missing-in-leadership-development

Moldoveanu, M., & Narayandas, D. (2019). The Future of Leadership Development (pp. 40-48). *Harvard Business Review*, March-April 2019.

Moreno-Jiménez, E., Flor-García, M., Terreros-Roncal, J., Rábano, A., Cafini, F., & Pallas-Bazarra, N. et al. (2019). Adult hippocampal neurogenesis is abundant in neurologically healthy subjects and drops sharply in patients with Alzheimer's disease. *Nature Medicine, 25*(4), 554-560. doi: 10.1038/s41591-019-0375-9

NASA, US Government. Climate Change: How do we know?. Retrieved from https://climate.nasa.gov/evidence/

New Scientist. (2017). The World Urgently Needs Critical Thinking, Not Gut Feeling. Retrieved from https://www.newscientist.com/article/mg23631563-900-the-world-urgently-needs-critical-thinking-not-gut-feeling/

New Scientist. (2019). Brain mysteries: A user's guide to the biggest questions of the mind. Retrieved 23 June 2019, from https://www.newscientist.com/article/ mg24232350-800-brain-mysteries-a-users-guide-to-the-biggest-questions-of-the-mind/

Nielsen, J., Zielinski, B., Ferguson, M., Lainhart, J., & Anderson, J. (2013). An Evaluation of the Left-Brain vs. Right-Brain Hypothesis with Resting State Functional Connectivity Magnetic Resonance Imaging. *Plos ONE, 8*(8), e71275. doi: 10.1371/ journal.pone.0071275

Northouse, P. (2016). *Leadership: Theory and Practice* (7th ed.). Los Angeles: SAGE.

O'Broin, A., & Palmer, S. (2019). Chapter 35 The coaching relationship: A key role in coaching processes and outcomes. In S. Palmer & A. Whybrow, *Handbook of Coaching Psychology: A Guide for Practitioners* (2nd ed., pp. 471-486). New York: Routledge.

Odom, T. (2015). Scientists Make Better Leaders: Using Ideas for the Common Good. [Blog]. Retrieved from https://www.huffpost.com/entry/scientists-make-better-leaders_b_7655784

Our World in Data. (2019). Retrieved 7 July 2019, from https://ourworldindata.org/

Palmer, S., & Whybrow, A. (2019). *Handbook of Coaching Psychology: A Guide for Practitioners* (2nd ed., pp. p. 51-67, 256-269, 324-340, 358-368). New York, NY: Routledge.

Panchal, S., Palmer, S., & Green, S. (2019). From Positive Psychology to the Development of Positive Coaching. In S. Palmer & A. Whybrow, *Handbook of Coaching Psychology: A guide for practitioners* (2nd ed., pp. 51-67). New York, NY: Routledge.

Patterson, K., Grenny, J., & McMillan, R. (2002). *Crucial Conversations*. Blacklick, USA: McGraw-Hill Professional Publishing.

Pearse, C. (2018). 5 Reasons Why Leadership Is In Crisis. [Blog]. Retrieved from https://www.forbes.com/sites/chrispearse/2018/11/07/5-reasons-why-leadership-is-in-crisis/#2b5c53b23aca

Peltier, B. (2010). *The Psychology of Executive Coaching* (2nd ed., pp.1-31, 59-79, 137-174, 175-209). New York, NY: Taylor and Francis.

Petrikowski, N. P. (2019). Angela Merkel: Chancellor of Germany. In *Encyclopaedia Brittanica*. Retrieved from https://www.britannica.com/biography/Angela-Merkel

Phillips, J. (2007). Measuring the ROI of a Coaching Intervention, Part 2. *Performance Improvement*, 46(10), 10-23. doi: 10.1002/pfi.167

Pinker, S. (2019). *Enlightenment Now: The Case for Reason, Science, Humanism, and Progress*. London: Penguin Books Ltd.

Possert, J. (2017). The Three Most Relevant Stages of Human Development in the Modernized World. [Blog]. Retrieved from https://libraryofconcepts.wordpress.com/2017/10/29/three-most-relevant-stages-of-human-development-nowadays-kegan-13/

Powers, A. (2017). These Skills Will Make You A Great CEO. [Blog]. Retrieved from https://www.forbes.com/sites/annapowers/2017/07/28/these-skills-will-make-you-a-great-ceo/#646374b2387a

Preston, J., Ritter, R., & Hepler, J. (2013). Neuroscience and the Soul: Competing explanations for the human experience. *Cognition*, 127(1), 31-37. doi: 10.1016/j.cognition.2012.12.003

Prinstein, M. (2017). *Popular: The Power of Likability in a Status-Obesessed World*. New York, NY: Penguin Random House LLC.

Ragonese-Campbell, R. (2017). Benefits of Executive Coaching. Retrieved 8 July 2019 from https://www.proveritas.com.au/coaching

Ragonese-Campbell, R. (2017). How to Achieve Career Goals Successfully: 4 Steps from Coaching Science. [Blog]. https://www.proveritas.com.au/blog-home/how-to-achieve-career-goals-successfully

Ragonese-Campbell, R. (2017). Should You Include Mindfulness in Leadership Development? Yes, but it Depends …. [Blog]. https://www.proveritas.com.au/blog-home/should-you-include-mindfulness-in-leadership-development-yes-but-it-depends

Ragonese-Campbell, R. (2017). What to Look For In a Coach – Start With These 3 Questions. [Blog]. Retrieved from https://www.proveritas.com.au/blog-home/what-to-look-for-in-a-coach-start-with-these-3-questions

Ragonese-Campbell, R. (2018). 10 Leadership lessons from International Coaching in Leadership Forum – Harvard Medical School. [Blog]. https://www.proveritas.com.au/blog-home/10-leadership-lessons-from-international-coaching-in-leadership-forum-harvard-medical-school

Ragonese-Campbell, R. (2018). Don't be a Burnout Statistic – it's time to renew. [Blog]. Retrieved from https://www.proveritas.com.au/blog-home/don-t-be-a-burnout-statistic-it-s-time-to-renew

Ragonese-Campbell, R. (2018). Nexus of Neuroscience, Coaching and Leadership: next frontier. [Blog]. Retrieved from https://www.proveritas.com.au/blog-home/nexus-of-neuroscience-coaching-and-leadership-next-frontier

Rainie, L. (2019). Trust and Distrust in America. [Blog]. Retrieved from https://www.people-press.org/2019/07/22/trust-and-distrust-in-america/

Rekalde, I., Landeta, J., Albizu, E., & Fernandez-Ferrin, P. (2017). Is executive coaching more effective than other management training and development methods?. *Management Decision*, 55(10), 2149-2162. doi: 10.1108/md-10-2016-0688

Return on Investment in Executive Coaching: Effective Organizational Change. (2012). *Development And Learning In Organizations: An International Journal*, 26(5). doi: 10.1108/dlo.2012.08126eaa.009

Reynolds, A. (2017). Teams Solve Problems Faster When They're More Cognitively Diverse. [Blog]. Retrieved from https://hbr.org/2017/03/teams-solve-problems-faster-when-theyre-more-cognitively-diverse?autocomplete=true

Rigby, C., & Ryan, R. (2018). Self-Determination Theory in Human Resource Development: New Directions and Practical Considerations. *Advances In Developing Human Resources*, 20(2), 133-147. doi: 10.1177/1523422318756954

Ritchie, Roser, Mispy, & Ortiz-Ospina. (2018). Measuring Progress Towards the Sustainable Development Goals – SDG Tracker. (2019). Retrieved 7 July 2019, from https://sdg-tracker.org/

Roberts, B., Luo, J., Briley, D., Chow, P., Su, R., & Hill, P. (2017). A Systematic Review of Personality Trait Change Through Intervention. *Psychological Bulletin*, 143(2), 117-141. doi: 10.1037/bul0000088

Rohrer, J., Egloff, B., & Schmukle, S. (2015). Examining the Effects of Birth Order on Personality. *SSRN Electronic Journal*. doi: 10.2139/ssrn.2704310

Rosling, H. (2018). *Factfulness: Ten Reasons We're Wrong About The World – And Why Things Are Better Than You Think*. London: Hodder & Stoughton General Division.

Runciman, D. (2013). Rat-a-tat-a-tat-a-tat-a-tat. Retrieved from https://www.lrb.co.uk/v35/n11/david-runciman/rat-a-tat-a-tat-a-tat-a-tat

Runciman, D. (2018). Why Replacing Politicians with Experts is a Reckless Idea. Retrieved from https://www.theguardian.com/news/2018/may/01/why-replacing-politicians-with-experts-is-a-reckless-idea

Ryan, R., & Deci, E. (2017). *Self-Determination Theory: Basic Psychological Needs in Motivation, Development, and Wellness* (1st ed., pp. 239-271, 382-400, 532-558). New York: Guilford Press.

Ryan, R., Soenens, B., & Vansteenkiste, M. (2018). Reflections on Self-determination Theory as an Organizing Framework for Personality Psychology: Interfaces, integrations, issues, and unfinished business. *Journal Of Personality*, 87(1), 115-145. doi: 10.1111/jopy.12440

Ryan, W., & Ryan, R. (2019). Toward a Social Psychology of Authenticity: Exploring Within-Person Variation in Autonomy, Congruence, and Genuineness Using Self-Determination Theory. *Review Of General Psychology*, 23(1), 99-112. doi: 10.1037/gpr0000162

Sainsbury, M. (2019). Comment: A Closer Look at Scott Morrison's CV. [Blog]. Retrieved from https://www.msn.com/en-au/news/australia/comment-a-closer-look-at-scott-morrisons-cv/ar-BBTqoLg

Salovey, P. (2018). We Should Teach All Students, in Every Discipline, to Think Like Scientists. *Scientific American*, (June 2018). Retrieved from https://www.scientificamerican.com/article/we-should-teach-all-students-in-every-discipline-to-think-like-scientists/

Sandberg, S. (2015). *Lean In*. Random House UK.

Sato, W. (2016). Scientists' Personality, Values, and Well-being. *Springerplus*, 5(1). doi: 10.1186/s40064-016-2225-2

Schindler, J. (2018). The Benefits of Cognitive Diversity [Blog]. Retrieved from https://www.forbes.com/sites/forbescoachescouncil/2018/11/26/the-benefits-of-cognitive-diversity/#71eedf625f8b

Schwab, K. (2016). The Fourth Industrial Revolution: What it means, how to respond [Blog]. Retrieved from https://www.weforum.org/agenda/2016/01/the-fourth-industrial-revolution-what-it-means-and-how-to-respond/

Schwab, K. (2018). The Fourth Industrial Revolution. (2018). [Blog]. Retrieved from https://www.britannica.com/topic/The-Fourth-Industrial-Revolution-2119734/media/1/2119734/229695

Science Rising. (2019). Retrieved 1 July 2019, from https://www.sciencerising.org/

Shah, P., Michal, A., Ibrahim, A., & Rhodes, R. (2017). What Makes Everyday Scientific Reasoning So Challenging?. In B. Ross, *Psychology of Learning and Motivation* (pp. 251-290). Elsevier Inc. Retrieved from https://doi.org/10.1016/bs.plm.2016.11.006

Shashkevich, A. (2019). When Are You Most Likely to Catch Other People's Emotions? [Blog]. Retrieved from https://greatergood.berkeley.edu/article/item/when_are_you_most_likely_to_catch_other_peoples_emotions

Shepherd, M. (2018). Stop Using The Word 'Nerd' – The Future Of Science May Depend On It [Blog]. Retrieved from https://www.forbes.com/sites/marshallshepherd/2018/04/30/stop-using-the-word-nerd-the-future-of-science-may-depend-on-it/#396b39c25dbc

Shmerling, MD, R. (2017). Right Brain/Left Brain, Right? [Blog]. Retrieved from https://www.health.harvard.edu/blog/right-brainleft-brain-right-2017082512222

Siersdorfer, D. (2019). We have the Tools to Beat Climate Change. Now we need to Legislate [Blog]. Retrieved from https://www.weforum.org/agenda/2019/04/we-have-the-tools-to-beat-climate-change-now-we-need-the-right-laws/

Snowden, D., & Boone, M. (2007). A Leader's Framework for Decision Making. *Harvard Business Review*. Retrieved from https://hbr.org/2007/11/a-leaders-framework-for-decision-making

Southwick, S., Bonanno, G., Masten, A., Panter-Brick, C., & Yehuda, R. (2014). Resilience Definitions, Theory, and Challenges: Interdisciplinary perspectives. *European Journal Of Psychotraumatology*, 5(1), 25338. doi: 10.3402/ejpt.v5.25338

Spears, R., & Tausch, N. (2012). Chapter 14: Prejudice and Intergroup Relations. In M. Hewstone, W. Stroebe & K. Jonas, *An Introduction to Social Psychology* (5th ed., pp. 451-498). Glasgow: BPS Blackwell.

Spence, G., & Deci, E. (2013). Self-determination Theory Within Coaching Contexts: Supporting Motives and Goals that Promote Optimal Functioning and Well-being. In S. David, D. Clutterbuck & D. Megginson, *Beyond Goals: Effective Strategies for Coaching and Mentoring* (1st ed., pp. 85-108). Surrey, England: Gower Publishing.

Stefon, M. Francis: Pope [Blog]. Retrieved from https://www.britannica.com/biography/Francis-I-pope

Standards Australia Limited. (2011). *Coaching in Organizations*. Sydney: SAI Global Ltd.

Stelter, R. (2015). *Guide to Third Generation Coaching*. Springer.

Tan, C. (2012). *Search Inside Yourself: The Unexpected Path to Achieving Success, Happiness (and World Peace)*. New York, NY: HarperCollins Publisher.

Tashiro, T. (2017). *Awkward: The Science of Why We're Socially Awkward and Why That's Awesome*. New York, NY: HarperCollins Publishers.

Technocracy. Retrieved 25 July 2019, from https://en.wikipedia.org/wiki/Technocracy

The Best-Performing CEOs in the World. (2019). [Blog]. Retrieved from https://hbr.org/2018/11/the-best-performing-ceos-in-the-world-2018

The bottom line of executive coaching. (2006). *Development And Learning In Organizations: An International Journal*, *20*(6), 32-34. doi: 10.1108/14777280610706220

The Imagination Institute – The Imagination Institute is dedicated to making progress on the measurement, growth, and improvement of imagination across all sectors of society. Retrieved 30 July 2019, from http://www.imagination-institute.org/

The Oxford Review. (2018). *The Essential Guide to Evidence-Based Practice*. Retrieved from https://www.oxford-review.com/evidence-based-practice-essential-guide/

The Oxford Review. (2018). *There Are Too Many Leadership Concepts...But Which Ones Are Redundant?*. Oxford, UK: Oxford Review Enterprises Ltd.

The Oxford Review. (2018). *Using Evidence In Evidence-Based Practice: Integrating Practice And Research* (pp. 17-22). Oxford Review Enterprises Ltd.

The Oxford Review. (2018). *What Does The Current Research Say About The Characteristics of Successful CEOs – Special Report*. Oxford Review Enterprises Ltd.

The Oxford Review. (2019). *Just how valid is the idea of emotional intelligence?*. Oxford, UK: Oxford Review Enterprises Ltd.

The Path to Becoming a CEO. [Blog]. Retrieved from https://www.investopedia.com/articles/financialcareers/08/ceo-chief-executive-career.asp

The Ten Principles. (2019). Retrieved 24 July 2019, from https://www.unglobalcompact.org/what-is-gc/mission/principles

The Treasury, Australian Government. (2019). *Final Report of the Royal Commission into Misconduct in the Banking, Superannuation and Financial Services Industry*. ACT: Australian Government. Retrieved from https://treasury.gov.au/publication/p2019-fsrc-final-report

Thomson, H. (2019). Sibling rivalry: How birth order affects your personality and health. *New Scientist*. Retrieved from https://www.newscientist.com/article/mg24332391-800-sibling-rivalry-how-birth-order-affects-your-personality-and-health/

Tikhenkaia, O. (2018). *CEO characteristics and innovations: an empirical study of European public companies of pharmaceutical industry* (Master of Corporate Finance). St Petersburg University, Graduate School of Management.

Tsusaka, M., Greiser, C., Krentz, M., & Reeves, M. (2019). *The Business Imperative of Diversity*. BCG Henderson Institute, Boston Consulting Group. Retrieved from https://www.bcg.com/publications/2019/winning-the-20s-business-imperative-of-diversity.aspx

United Nations Educational, Scientific and Cultural Organization (UNESCO). (2015). *UNESCO Science Report*. Paris: United Nations Educational, Scientific and Cultural Organization. Retrieved from http://uis.unesco.org/sites/default/files/documents/unesco-science-report-towards-2030-part1.pdf

United Nations. (2018). *The International Panel on Climate Change, Summary for Policymakers*. IPCC. Retrieved from https://www.ipcc.ch/sr15/

United Nations. (2019). About the Sustainable Development Goals. Retrieved 23 June 2019, from https://www.un.org/sustainabledevelopment/sustainable-development-goals/

University of Oxford. [Blog]. Retrieved from https://ophi.org.uk/policy/national-policy/gross-national-happiness-index/

Wallenfelt, J. (2019). Scott Morrison, Prime Minister Of Australia. Retrieved from https://www.britannica.com/biography/Scott-Morrison

Wang, X., Kim, T., & Lee, D. (2016). Cognitive diversity and team creativity: Effects of team intrinsic motivation and transformational leadership. *Journal Of Business Research*, 69(9), 3231-3239. doi: 10.1016/j.jbusres.2016.02.026

Wasylyshyn, K. (2017). From here to certainty: Becoming CEO and how a trusted leadership advisor (TLA) helped the client get there. *Consulting Psychology Journal: Practice And Research*, 69(1), 1-25. doi: 10.1037/cpb0000071

Weintraub, K. (2019). The Adult Brain Does Grow New Neurons After All, Study Says. *Scientific American*. Retrieved from https://www.scientificamerican.com/article/the-adult-brain-does-grow-new-neurons-after-all-study-says/?print=true

Weir, K. (2012). The Pain of Social Rejection. *American Psychological Association*, (Vol 43, No. 4), 50. Retrieved from https://www.apa.org/monitor/2012/04/rejection

Weller, C. (2015). How the Patron Saint of Introverts is Quietly Revolutionising American Culture. Retrieved from https://www.businessinsider.com.au/susan-cains-quiet-revolution-is-changing-america-2015-8

Who, What, Why: What can technocrats achieve that politicians can't?. (2011). [Blog]. Retrieved from https://www.bbc.com/news/magazine-15720438

Why Cognitive Diversity is the Latest Issue Preoccupying Hiring Managers. (2018). [Blog]. Retrieved from https://www.managers.org.uk/insights/news/2018/october/why-cognitive-diversity-is-the-latest-issue-preoccupying-hiring-managers

Wilkinson, D. (2018). *Executive Coaching and its Outcomes: What the research actually says*. Oxford: The Oxford Review.

Wilkinson, D. (2018). *Leader Differences And Their Impact On Leadership Outcomes*. Oxford: The Oxford Review.

Wilkinson, D. (2019). Just How Valid is the Idea of Emotional Intelligence?. *The Oxford Review*. Retrieved from https://www.oxford-review.com/emotional-intelligence-validation/

Williams, J., & Penman, D. (2014). *Mindfulness: A practical guide to Finding Peace in a Frantic World*. London: Piatkus.

Williams, R. (2016). Four Scientists Who Became World Leaders [Blog]. Retrieved from https://www.abc.net.au/news/2016-11-07/four-scientists-who-became-world-leaders/7987212

Williams, T. (2019). America's Top CEOs and Their College Degrees [Blog]. Retrieved from https://www.investopedia.com/articles/professionals/102015/americas-top-ceos-and-their-college-degrees.asp

Wilson, C. (2019). Science's dark secret is out. *New Scientist*, (3222), 7.

Winterhalter, B. (2014). ISTJ? ENFP? Careers Hinge on A Dubious Personality Test [Blog]. Retrieved from https://www.bostonglobe.com/opinion/2014/08/30/istj-enfp-careers-hinge-dubious-personality-test/8ptUGXhu6DndFdjCngcxSN/story.html%20

World Economic Forum. (2018). *The Future of Jobs Report*. Geneva, Switzerland: World Economic Forum. Retrieved from https://www.weforum.org/reports/the-future-of-jobs-report-2018

World Economic Forum. (2019). *The Global Risks Report 2019, 14th Edition*. Geneva, Switzerland: World Economic Forum. Retrieved from https://www.weforum.org/reports/the-global-risks-report-2019

World Health Organization (WHO). (2019). *Burn-out an "Occupational Phenomenon": International Classification of Diseases*. WHO. Retrieved from https://www.who.int/mental_health/evidence/burn-out/en/

Wright, P. (2019). Psychologists say Personality is all About the "Big Five" Traits – what are they? [Blog]. Retrieved from https://www.abc.net.au/life/big-five-personality-traits-backed-by-science-explained/10749608

Zelenski, J., Santoro, M., & Whelan, D. (2012). Would Introverts be Better Off if they Acted More Like Extraverts? Exploring emotional and cognitive consequences of counterdispositional behavior. *Emotion*, *12*(2), 290-303. doi: 10.1037/a0025169

Zenger, J., & Folkman, J. (2016). The Trickle-Down Effect of Good (and Bad) Leadership. *Harvard Business Review*. Retrieved from https://hbr.org/2016/01/the-trickle-down-effect-of-good-and-bad-leadership

www.ingramcontent.com/pod-product-compliance
Lightning Source LLC
Chambersburg PA
CBHW071203210326
41597CB00016B/1651